Designing Portals: Opportunities and Challenges

Ali Jafari
Indiana University Purdue University Indianapolis (IUPUI), USA

Mark Sheehan
Montana State University, USA

Information Science Publishing
Hershey • London • Melbourne • Singapore • Beijing

Acquisitions Editor:	Mehdi Khosrow-Pour
Senior Managing Editor:	Jan Travers
Managing Editor:	Amanda Appicello
Development Editor:	Michele Rossi
Copy Editor:	Maria Boyer
Typesetter:	Tamara Gillis
Cover Design:	Michelle Waters
Printed at:	Integrated Book Technology

Published in the United States of America by
    Information Science Publishing (an imprint of Idea Group Inc.)
    701 E. Chocolate Avenue, Suite 200
    Hershey PA 17033
    Tel: 717-533-8845
    Fax: 717-533-8661
    E-mail: cust@idea-group.com
    Web site: http://www.idea-group.com

and in the United Kingdom by
    Information Science Publishing (an imprint of Idea Group Inc.)
    3 Henrietta Street
    Covent Garden
    London WC2E 8LU
    Tel: 44 20 7240 0856
    Fax: 44 20 7379 3313
    Web site: http://www.eurospan.co.uk

Copyright © 2003 by Idea Group Inc. All rights reserved. No part of this book may be reproduced in any form or by any means, electronic or mechanical, including photocopying, without written permission from the publisher.

Library of Congress Cataloging-in-Publication Data

Designing portals: opportunities and challenges / [edited by] Ali Jafari, Mark Sheehan.
    p. cm.
  Includes bibliographical references and index.
  ISBN 1-59140-108-9 (hardcover)
  1. Internet in higher education. 2. Web portals--Design. 3. Education technology. 4. Information technology. I. Jafari, Ali. II. Sheehan, Mark (Mark C.)

LB1044.87 .D49 2002
025.04--dc21
                                              2002027312

eISBN 1-59140-116-X

British Cataloguing in Publication Data
A Cataloguing in Publication record for this book is available from the British Library.

NEW Titles from Information Science Publishing

- **Web-Based Education: Learning from Experience**
  Anil Aggarwal
  ISBN: 1-59140-102-X: eISBN 1-59140-110-0, © 2003
- **The Knowledge Medium: Designing Effective Computer-Based Learning Environments**
  Gary A. Berg
  ISBN: 1-59140-103-8; eISBN 1-59140-111-9, © 2003
- **Socio-Technical and Human Cognition Elements of Information Systems**
  Steve Clarke, Elayne Coakes, M. Gordon Hunter and Andrew Wenn
  ISBN: 1-59140-104-6; eISBN 1-59140-112-7, © 2003
- **Usability Evaluation of Online Learning Programs**
  Claude Ghaoui
  ISBN: 1-59140-105-4; eISBN 1-59140-113-5, © 2003
- **Building a Virtual Library**
  Ardis Hanson & Bruce Lubotsky Levin
  ISBN: 1-59140-106-2; eISBN 1-59140-114-3, © 2003
- **Design and Implementation of Web-Enabled Teaching Tools**
  Mary F. Hricko
  ISBN: 1-59140-107-0; eISBN 1-59140-115-1, © 2003
- **Designing Portals: Opportunities and Challenges**
  Ali Jafari and Mark Sheehan
  ISBN: 1-59140-108-9; eISBN 1-59140-116-X, © 2003
- **Challenges of Teaching with Technology Across the Curriculum: Issues and Solutions**
  Lawrence A. Tomei
  ISBN: 1-59140-109-7; eISBN 1-59140-117-8, © 2003

Excellent additions to your institution's library! Recommend these titles to your librarian!

To receive a copy of the Idea Group Inc. catalog, please contact (toll free) 1/800-345-4332, fax 1/717-533-8661, or visit the IGI Online Bookstore at: http://www.idea-group.com!

Note: All IGI books are also available as ebooks on netlibrary.com as well as other ebook sources. Contact Ms. Carrie Stull at <cstull@idea-group.com> to receive a complete list of sources where you can obtain ebook information or IGP titles.

Designing Portals: Opportunities and Challenges

Table of Contents

Preface .... **vi**
    Ali Jafari, IUPUI, USA

Chapter I. Introduction **1**
    Mark Sheehan, Montana State University, USA
    Ali Jafari, IUPUI, USA

SECTION I: DESIGNING PORTALS: THEORY AND PRACTICE

Chapter II. The ABCs of Designing Campus Portals **7**
    Ali Jafari, IUPUI, USA

Chapter III. Keeping Your Eyes on the Prize: Using Inquiry to Increase the Benefits of Institutional Portals **28**
    Stephen C. Ehrmann, The Flashlight Program, USA

Chapter IV. Portals: Your Institution's Reputation Depends on Them **37**
    William H. Graves and Kirsten Hale, Eduprise, USA

Chapter V. Developing A Portal Channel Strategy **51**
    Jameson Watkins, University of Kansas Medical Center, USA

Chapter VI. Campus Portal Strategies **68**
    David L. Eisler, Weber State University, USA

Chapter VII. The Next Generation of Internet Portals **89**
    Ali Jafari, IUPUI, USA

SECTION II: CASE STUDIES OF CAMPUS PORTALS

Chapter VIII. Indiana University's Enterprise Portal as a Service Delivery Framework 102
    James Thomas, Indiana University, USA

Chapter IX. Begin with the End (User) in Mind: Planning for the San Diego State University Campus Portal 127
    James P. Frazee, Rebecca Vaughan Frazee and David Sharpe, San Diego State University, USA

Chapter X. Values-Based Design of Learning Portals as New Academic Spaces 162
    Katy Campbell and Robert Aucoin, University of Alberta, Canada

Chapter XI. Building a Campus Portal—A Strategy that Succeeded 186
    Anne Yandell Bishop, Wake Forest University, USA

SECTION III: VENDORS' PERSPECTIVES

Chapter XII. Portals Unlock the Knowledge that Drives Business Value 202
    Robert Duffner, BEA Systems, Inc., USA

Chapter XIII. Portal Technology and Architecture: Past, Present and Future 220
    Christopher Etesse, Blackboard Inc., USA

Chapter XIV. Building a Virtual Campus 238
    Stephen Ast and Cassandra Gerfen, eCollege, USA

Appendix I. Online Survey Results 256
    Mark Sheehan, Montana State University, USA

Appendix II. Educational Portal White Paper 270
    Ali Jafari, IUPUI, USA

About the Authors 291

Index 298

Preface

At the AAHE, American Association of Higher Education, Summer Institute on Teaching, Learning, and Technology in the summer of 1997, William M. Plater, the Executive Vice Chancellor and Dean of the Faculties at Indiana University Purdue University Indianapolis, IUPUI, defined his vision to present all of the IUPUI course syllabi on the Web. His request cultivated my thinking toward the conceptualization and design of Internet portals and planted the seed for my first portal project. The initial solution proposed to Dean Plater was the development of a new personal and dynamic Web environment. This dynamic environment requires that every student and instructor automatically receives access to some teaching and learning tools upon authentication through a single Website. Such methods of packaging classroom resources and tools into a single, centrally managed Web environment are now known as course management software (CMS) systems. Dynamic, role-based Web environments tailored specifically to selected groups of users (or members of an institution) are now known as Internet portals.

At that time, I was the Director of the WebLab and Associate Professor of Purdue School of Engineering and Technology at IUPUI. WebLab was a research and development laboratory initiated to explore and develop new Web-based educational technology solutions for the university. I was working with Amy Conrad Warner, the Executive Director of Community Learning Network to develop one of the very first "Web-based" distance-learning courses at IUPUI. Our initial beta-test environment included redesigning an existing video-based introductory Chemistry course into a Web-based course. Establishing a defined set of functional requirements enabled us to develop a tool set that would not only meet the needs of Chemistry 101 instructors and learners, but serve virtually 100% of the courses offered on the IUPUI campus and throughout the Indiana University enterprise. In less than six months, I assembled a team of enterprising innovative students, including an undergraduate student who had developed an online testing software solution that would become part of our tool set. Together, David Mills and other students working in the WebLab developed a complete course management system loaded with a message board, chat rooms, email and many other tools. We called the system Oncourse.

Linked to the university student enrollment database, Oncourse holds the distinction of being one the first enterprise course management portals implemented at an educational institution (Jafari, 1999, 2000). Today, Oncourse serves all eight Indiana University campuses, dynamically creating a course management site for every course being offered in the university. Oncourse remains a good example of a system that is both dynamic and enterprise-wide. Dynamic because it automatically enables and disables students' and faculty access to courses and other resources based on the course registration data which resides in the university databases. Enterprise-wide because it offers dynamic services to the entire population of the university through direct connectivity to the university database systems providing up-to-date access to relevant course enrollment data.

As the principal architect of the Oncourse learning environment design, I assumed many roles in the development of the project. I played the role of a conceptual thinker and architect to invent, design and sell a new complementary environment for teaching and learning. Recall that in 1997, the notion of CMS was very new and only a small portion of faculty members had a working knowledge of the capacity of the Web and its applications in teaching and learning. Therefore, my biggest challenge was to sell the concept of the Web as a new useful teaching and learning tool, and to articulate how this new technology would revolutionize information management while fueling learning on demand.

To launch the concept, the environment must be easy to use, require little or no training and enable faculty members to learn at their own rate. Therefore, the top three functional requirements became: ease of use, ease of use and ease of use. Oncourse offered new Web-based tools and resources that made it very sticky—the stickiness would invite learners back time and time again for current up-to-date information otherwise not available to them seven days a week and 24 hours a day. Among the faculty, the early adopters of Oncourse began to instantaneously introduce the concepts of distance learning and Web access into their classroom teaching environment. With faculty embracing the technology, Oncourse provided a vehicle through which I could define the distinct advantages and need for portals in educational institutions.

The Oncourse navigation system was conceptualized in much the same manner a typical portal environment is conceptualized today. All users--students and course assistants and faculty--go to a single website, http://oncourse.iu.edu. Each user is authenticated into the Oncourse environment using the same university network ID required to access e-mail and other campus-wide IT services. Students and faculty depend upon their Network ID to conduct a number of university transactions, so an additional unique user ID need not be established. Once users enter their username and password, they automatically receive an updated list of their registered courses on the following page. Students view a list of courses in which they are currently enrolled and faculty view a list of courses that they are assigned to teach. Each course listing appears as a hyperlink taking users to the course management portion of Oncourse. Oncourse offers dynamic role-based services. For instance, the faculty member of record automatically receives authoring privilege to create and edit syllabi, course contents, etc., but these authoring rights are automatically blocked from student users. With this notion, we managed to create a portal environment offering dynamic and role-based services to the entire population of the university.

Consistent with the functional requirements for ease of use, I was strongly convinced to offer a fixed template interface instead of letting each faculty member design his/her own course management Website. This offered two major important roles in making the Oncourse project a success: the usability advantages or ease of use and little or no additional investment in user-support services (i.e., helpdesk). Not permitting faculty members to design their own course template created a comprehensive branding feature providing a consistent student-centered user interface. Once a student learned the navigational and user interface of a course, he/she can apply the learning toward other courses created by other faculty members. Second, I was not convinced that all faculty members knew about the fundamental design requirements of creating a quality user interface. Additionally, eliminating the opportunity to create a new template for each course, faculty could focus their innovations on learning objectives rather than tinkering in the world of user interface design and navigational differences that would detract from students' ability to focus on learning. Students, for instance, may not easily find the location of the syllabus, message boards and other resources if each course were developed by a different faculty member with a different learning style. Having more than one template would also complicate central support services delivery or reduce the complexity of providing helpdesk services in a timely manner. With this notion, Oncourse offered a fixed course management template with a fixed menu including categories for Syllabus, Lessons, In-Touch, Tools and Help. The notion of using a fixed course template was later offered by commercial course management systems.

Technically, the course portal was designed as an enterprise system to offer services to all campuses of Indiana University with little or no customization required. With this notion, certain design principles had to be selected while preserving technical requirements such as scalability, performance, load-balancing, integration and maintenance.

After handing over the Oncourse project from the R&D environment of the WebLab to the University Information Technology Services for the system-wide implementation in 1999, I began my next portal project called ANGEL, A New Global Environment for Learning. With seed funding received from the School of Engineering and Technology, and a commitment from David Mills, the lead developer from the Oncourse Team, we were able to further develop ANGEL in a new research and development laboratory called CyberLab located at the IUPUI campus. In contrast to Oncourse which was hard coded to work with the information technology framework of the university, ANGEL was designed to work with any system, to be easy to install and integrate with any infrastructure in any school. From the beginning, ANGEL was designed as a modular system, offering new features to enhance the portal environment. Additionally, the modular capabilities of ANGEL offered the feature of expandability and performance requirements of portals since various portals' tasks and services can be distributed among different servers.

In 1999, through some collaborative research with a colleague at Florida State University, I became increasingly interested in the conceptualization and design of intelligent agents to address teaching and learning needs. My interest intensified as I noticed that the teaching and learning environments, more specifically the CMS and campus portals, became more labor intensive to maintain while advances in technology continued to make portals easier to use at an exponential rate. Faculty colleagues who were teaching online courses, for instance, indicated that they were spending more time teaching an online course than teaching the same course in the traditional classroom lecture setting. While the increased time commitment required to engage learners at a distance has nothing to do with design of user interface or ease of use aspects of the environment, it has everything to do with the magnitude of tasks users were required to perform. There were many logistical matters and maintenance requirements in a Web environment that affects its ease of use. It became very clear to me that current CMS and portal technologies are "dumb," and are not designed to offer intelligent services. With this, I quickly saw the multitude of applications for intelligent agents in teaching and learning environments. Conceptually, the intelligent agents can act like a human agent offering personal services to users of a portal. Technically speaking, the intelligent agents can be integrated into a portal or CMS software environment to accept certain responsibilities and to perform certain tasks on behalf of its uses. The ANGEL environment from the ground up was designed as an agent-based portal environment where the third-party vendors or end-users' institutions can design and integrate Intelligent Agents into the ANGEL portal environment.

ANGEL was certainly another successful project. With financial support received from Indiana University Advance Research Technology Institute (ARTI), a small company was formed to commercialize the ANGEL CMS and portal software environment. In July of 2000, ANGEL was transferred from my academic IUPUI CyberLab into the newly formed company, CyberLearning Labs Inc. This migration enabled me to return to my passion to explore and develop new technology innovations.

In late 2000 right after my ANGEL project, I developed a white paper to conceptualize the design and development of an inter-campus educational portal to serve K-12 and higher education institutions (Jafari, 2001). The resulting paper, "Educational Portal White Paper," was submitted to the Indiana Higher Education Telecommunication System (IHETS). In contrast to a campus portal, which is meant to serve the community of a single campus, the educational portal is defined in my paper as a super portal environment, to be used by instructors and learners within a large number of educational institutions, such as all K-12 and higher ed institutions in a state or an entire nation. I saw tremendous value in the creation of a central educational portal environment that could be used for collaborative sharing of information, resources and learning objects among a statewide or national population of teachers and learners. For instance, a high school instructor developing a learning module for his chemistry class would be able to dynamically inform other chemistry teachers about his work, teachers who might be interested in integrating this module into their chemistry course. Similarly, the portal environment could offer opportunities for collaboration among learners with similar interests or similar learning disorders. State government could use this environment to offer teaching and learning resources to individual displaced workers, parochial schools and non-traditional learning providers. Anxious to build a strong workforce, state agencies can provide a powerful tool in attracting and retaining industry. The educational portal provides a single entry point to training and educational opportunities for the disenfranchised and often disengaged. Examples of resources included in the educational portal might include course management tools, state and community library resources, central file serving resources, and electronic portfolios. The educational portal was conceptualized as a profile-based intelligent portal environment using intelligent agents. The white paper explores many creative ideas for making the portal sticky, dynamic and easy to use. In the spring of 2002, IHETS received seed funding to further explore the educational portal project as a potential community service to educators and lifelong learners in the state of Indiana.

Besides my collaboration with Mark Sheehan writing this book in 2001, my attention was directed to a new R&D project. My third portal project provides yet another set of new requirements and interface design. At the time of writing this manuscript, this project does not have a given name. The project code name is DPP or Dynamic Personal Portal, being developed at the IUPUI CyberLab with collaboration with some other universities.

The DPP will invent a new interface and a new life-long teaching and learning portal environment for every learner. A learner may begin using the DPP environment from his or her freshman year in college, or perhaps attracted as high ability students even prior to high school graduation. The DPP environment follows students from high school to college, to graduate school and to their professional lives. The DPP offers many utilities including an electronic portfolio system that travels with students. It offers services like a personal home page (PHP), electronic portfolio and campus portal. It is conceptualized as a totally dynamic portal environment and offers a unique and life-long personal URL (Web address) to every student. The personal URL is based on the learner's email address. For instance if my email address is jafari@iupui.edu, my DPP address would be http://jafari.with.iupui.edu. Note the similarities between my email address and my personal URL. The only difference is the replacement of the "@" sign with a "with" word. This is logical, easy to use, easy to remember and enables learners to even make an educated guess to locate personal URLs for every member of an institution. If one knows my email address, he or she can guess my personal URL address. The "with" world within the domain name can be any word selected by an institution. The personal URL can stay with a student as a Web identity, letting him/her carry the "brand name" of his college throughout post-graduation professional and personal life (the inclusion of ".universityname.edu" in a personal URL). It would serve as the life-long personal URL that could appear on people's business cards.

As my new and current project, I am trying to further define, design, and develop the electronic portfolios system within the DPP framework through collaboration with other higher education institutions. In contrast with my Oncourse and ANGEL projects developed at IUPUI, the DPP and Electronic Portfolios will be designed and developed by a consortium of higher educations institutions and participating vendors. One of the most important requirements of DPP and Electronic Portfolios is the need for interoperability and transportability of learning accomplishments, therefore, it is very important that the DPP/Electronic Portfolios project be designed and accepted by more than one institution. With this notion, in late 2001, I initiated and founded the ePortConsortium. The DPP/Electronic Portfolios project is an open source initiative available to members of the consortium. The DPP framework holds a patent pending protection owned by Indiana University.

The more I reflect on our accomplishments and analyze emerging trends and opportunities for Internet portals, the more passionate I have become with respect to the development of intelligent portals for teaching and learning. We are in the infancy stages of conceptualizing and developing Internet portals, especially campus portals which optimize our teaching and learning needs. Every new day, large amounts of data, information and resources reside within the World Wide Web. We must continue to create the perfect user interface and Internet portal system that intelligently filters and provides mass customization of information and resources to serve learners on demand. Our next generation of portals must have the capacity to think, to learn, to reason and to maintain a certain level of autonomy.

Ali Jafari
Purdue School of Engineering and Technology, IUPUI

REFERENCES

Jafari, A. (1999). Putting everyone and every course on-line: The Oncourse environment. *WebNet Journal,* 1(4), 37-43.

Jafari, A. (1999). The rise of a new paradigm shift in teaching and learning. *T.H.E. Journal*, 27(3), 58-68.

Jafari, A. (2000). *Development of a New University-Wide Course Management System, Cases on Information Technology in Higher Education.* Hershey, PA: Idea Group Publishing.

Jafari, A. (2001). Educational Portal, *Proceedings of the International Conference on Intelligent Agents,* Las Vegas, USA, July.

ADDITIONAL RESOURCES

Website of ANGEL Course Management and Portal Software: *http://www.cyberlearninglabs.com/Products/*

Website of CyberLab at IUPUI: *http://cyberlab.iupui.edu/*

Website of CyberLearning Labs Inc.: *http://www.cyberlearninglabs.com/Products/*

Website of the Dynamic Personal Portals project: *http://with.iupui.edu/dpp.htm*

Website of the Electronic Portfolio Consortium (ePortConsortium): *http://www.eportconsortium.org/*

Website of the Indiana Higher Education Telecommunication System (IHETS): *http://www.ihets.org/*

Website of the Oncourse Project at Indiana University: *http://oncourse.iu.edu/*

ACKNOWLEDGMENTS

The authors wish to acknowledge Indiana University-Purdue University at Indianapolis and Montana State University for their support of the authors' work on this book. The authors thank their families for their patience and support during the project and thank the many colleagues and friends who have contributed to the development of their thinking about Internet portals. In particular, Mark Sheehan wishes to thank Keiko Pitter of Whitman College for the many discussions in which she helped frame his perspective.

In addition, the authors wish to thank the following individuals who participated in their Summer 2001 survey (in alphabetical order by institution).

Katherine Ranes, Arizona State University
Don Davis, Azusa Pacific University
Donald King, Jr., Ball State University
Stephen Pollard, California State University at Los Angeles
Kenneth Pflueger, California Lutheran University
Margaret Lynn Lester, Clarke College
Phil Lyles, Clemson University
Sissy Ehrhardt, College of Charleston
Janell K. Baran, Denison University
Peter Day, Emory University
Carole Meyers, Emory University
Brent T. Jaeger, Grinnell College
Blair Combs, Idaho State University
Kevin Shalla, Illinois Institute of Technology
Steve Brunner, Indiana University
Barry Walsh, Indiana University
Kirk Job Sluder, Indiana University
Gary Wiggins, Indiana University
Bill Orme, Indiana University Purdue University Indianapolis
Joy Starks, Indiana University Purdue University Indianapolis
Martha McCormick, Indiana University Purdue University Indianapolis
Ken Duckworth, Indiana University Purdue University Indianapolis
Ed Sullivan, Indiana University Purdue University Indianapolis
Edward C. Squires, Indiana University Purdue University Indianapolis
Sharon Hamilton, Indiana University Purdue University Indianapolis
Eugenia Fernandez, Indiana University Purdue University Indianapolis
Gloria Quiroz, Indiana University Purdue University Indianapolis
Ann Kratz, Indiana Indiana University Purdue University Indianapolis
Monica R. Peterson, Indiana University Purdue University Indianapolis

Amy C. Warner, Indiana University Purdue University Indianapolis
Tony Holderith, Interactive Business Solutions
Tim Foley, Lehigh University
Joseph Aulino, Northeastern Ohio Universities College of Medicine
Jerry Harder, Nazarene University
David Koehler, Princeton University
Judith Clark, Distributed National Electronic Resource (UK)
Stephen G. Landry, PhD, Seton Hall University
Beth Du Pont, Skidmore College
Morgan R. Olsen, Southern Methodist University
Sara Clark, Southwest Missouri State University
Steven Conway, Texas A&M University at Galveston
Fay Gibbons, The Australian National University
Nadine Stern, The College of New Jersey
Deirdre Zarrillo, The Sage Colleges
H. Leskinen, TTKK (Finland)
Katy Campbell, University of Alberta
Brian Alexander, University of California at Davis
Steve Terry, University of Memphis
Linda Miller, University of New Mexico
David McGuire, University of New Mexico
Tom Monaghan, University of Notre Dame
Gary Dobbins, University of Notre Dame
Joseph E. St. Sauver, PhD, University of Oregon
Linda H. Mantel, University of Portland
Ralph Pina, University of Stellenbosch (S. Africa)
Shan Evans, University of Texas at Austin
Amos Lakos, University of Waterloo
Ryan Schutt, Virginia Tech
Dave Eisler, Weber State University
John P. Jones, Wichita State University
Nicholas Rawlings, Yale University
Mark French, Zayed University (United Arab Emirates)

Chapter I

Introduction

Mark Sheehan
Montana State University, USA

Ali Jafari
IUPUI, USA

This is a book about Internet portals in higher education. It grew out of the editors' sense that the application of portal technologies to college and university needs is a much broader topic than can be addressed in a brief article or conference presentation.

Portals present unique strategic challenges in the academic environment. Their conceptualization and design requires the input of campus constituents who seldom interact and whose interests are often opposite. The implementation of a portal requires a coordination of applications and databases controlled by different campus units at a level that may never before have been attempted at the institution. Building a portal is as much about constructing intra-campus bridges as it is about user interfaces and content. Richard Katz (2000) sums it up concisely: "A portal strategy is difficult and perilous because many on campus are weary and suspicious of another new enterprise-wide information technology initiative, and because portals, by definition require across-the-institution agreements on approach and design that are hard to achieve in loosely coupled organizations like academic institutions."

So what *is* a portal? In the broad Internet context, definitions vary widely. The earliest portals to adopt the name, Yahoo! and Excite, both grew out of the Web search engine and Web index environments. Interestingly, Stanford University graduate students designed both.

The designers of Yahoo! wanted "a guide [to the Web]," "a list of favorites" and "a single place to find useful Websites" (Yahoo! Inc., 2002). When it was first released, Yahoo! quickly became the place to go to find an organized view of the explosively expanding universe of online information.

Copyright © 2003, Idea Group Inc. Copying or distributing in print or electronic forms without written permission of Idea Group Inc. is prohibited.

Excite began as a Web search engine, but in the mid-1990s, when it adopted its identity as a portal, it offered the first popular "personalizable" Web start page. With it, users could create their own points of entry into the broader Web by selecting from a variety of information options (which we would now call "channels") provided by Excite.

While it never adopted the name, portal, O'Reilly & Associates' Global Network Navigator (GNN) lays claim to being the first Web portal (O'Reilly, 2001). Introduced in 1993, the year before Yahoo! had its origins, GNN's features included GNN News, GNN Magazine, The Whole Internet Catalog, GNN Marketplace and the Navigator's Forum. It was later sold to America Online, and its basic concepts were incorporated into early versions of that service.

GNN was, and Yahoo! and Excite are outward-looking portals; they bring a measure of organization to the otherwise chaotic Internet and serve as an individual's point of entry into that vast information space. Eventually Yahoo! and Excite became Internet "destinations" in themselves, offering, in addition to Web navigation aids, a set of self-branded services to visitors. Yahoo! examples include Yahoo! Shopping and Yahoo! Personals. Excite, emphasizing its features for personalization, calls its services MyStocks, MyNews and so on.

As the concept of the Internet portal has evolved, though, portals have become more inward looking. In the commercial world, Amazon.com's is perhaps the perfect example: it provides the visitor with a customized view of everything Amazon sells. Similarly, campus portals, like the campus Websites that preceded them, are starting points for the exploration of campus resources. Most incorporate "feeds" of information from external sources: weather, national news and sports, and the like. But their real purpose is to draw the user into the campus Web space and from there into online aspects of the campus "living and learning" community.

Howard Strauss's (2000) early definition of portals in the higher education environment helps us distinguish portals from traditional Websites. It bears repetition and a bit of elaboration here. In Strauss's view, a true portal is:

- **Customized**—A true portal is a Web page whose format and information content are based on information about the user stored in the portal's database. When the user authenticates (logs in) to the portal, this information determines what the user will see.
- **Personalized**—The user can select and store a personal set of appearance and content characteristics for a true portal. These characteristics may be different for every user.

- **Adaptive**—The portal gets to "know" the user through information the user supplies and through information the portal is programmed to gather about the user. As the user's role in the institution changes (when a student becomes an employee, for example), a true portal will detect that change and adapt to it without human intervention.
- **Desktop-Oriented**—The goal of a portal is to mask the inner workings of the campus information systems from the user. Signing on to the portal keeps the user from having to sign onto each of the many systems, on-campus and off, that provide the portal content. The ultimate portal could become the user's point of entry not just into campus and Internet Web spaces, but also into his or her own desktop computer.

In Summer 2001, the editors of this book conducted an online survey to gather opinions about what a portal is perceived to be in the context of higher education. The results (campusportals, 2002) were presented in a poster session at the EDUCAUSE 2001 conference (Jafari & Sheehan, 2001). Table 1 summarizes the results. Clearly, the respondents accepted the ideas that a portal must be, in Strauss's terms, customized (questions 2, 4, 5, 11, 12 and 13), personalized (question 1), adaptive (question 7) and a potential replacement for the user's standard desktop environment (suggested, at least, by questions 8 and 9). Most respondents had high hopes that portals will incorporate "intelligent agent" features (question 13).

Respondents rejected the idea that a portal is "just" a Web page of links to other sites (question 3) and that a portal in the higher education context may not incorporate advertising (question 6), as most commercial portals do.

While 29% of respondents declined to speculate, 37% agreed that most schools and companies would *replace* their websites with portals by 2004 (question 14). Twenty-one percent thought it unlikely that this would occur before 2008; of these, 20% said, "never." The full survey results appear in the Appendix I of this book.

Of course defining portals is only the beginning. The rest of this book goes much further, describing in its three sections the current status of portals in higher education. Section 1, Designing Portals: Theory and Practice, provides insight into the role portals play in an institution's business and educational strategy. Section 2, Case Studies of Campus Portals, takes the reader through the processes of conceptualization, design and implementation of the portals (in different stages of development) at Indiana University, San Diego State University, the University of Alberta and Wake Forest University. Finally, Section 3, Vendors' Perspectives, offers insights from three producers of portal software systems in use at institutions of higher learning and elsewhere.

Copyright © 2003, Idea Group Inc. Copying or distributing in print or electronic forms without written permission of Idea Group Inc. is prohibited.

Table 1. Results of Online Survey

Question Number and Question	Strongly Agree	Agree	Neutral	Disagree	Strongly Disagree
1 To be considered a portal, a website must give me the option to customize it to look exactly the way I want it to.	25%	51%	9%	13%	1%
2 To be considered a portal, a website must greet me by name when I access it (after my initial login).	22%	35%	19%	22%	1%
3 "Portal" is just another word for a website that supplies links to many other pages.	6%	14%	5%	31%	45%
4 To be considered a portal, a website must accumulate information about my use of it AND it must customize accordingly its subsequent presentation of information to me.	13%	43%	21%	18%	6%
5 To be considered a portal, a website must be "intelligent," showing me only information and links that I often use.	9%	37%	19%	29%	6%
6 To be considered a portal, a website cannot display advertisements.	17%	11%	15%	48%	9%
7 To be considered a portal, a website must be linked to a database(s) of information about me and as my characteristics in the database change, the website must present different information to me.	26%	46%	17%	8%	3%
8 To be considered a portal, a website could be accessible from a personal digital assistant (e.g., a Palm Pilot).	9%	38%	25%	28%	0%
9 To be considered a portal, a website could be accessible from a Web-enabled mobile telephone.	11%	32%	29%	27%	2%
10 To be considered a portal, a website must function as an agent for me or as a personal assistant, helping me with my daily electronic communication needs.	19%	40%	19%	19%	1%
11 To be considered a portal, a website must require me to identify myself to it (i.e., authenticate with username and password or PIN) each time I access it.	38%	32%	8%	18%	5%
12 To be considered a portal, a website must allow people to use it without logging in.	4%	24%	16%	39%	16%
13 To be considered a portal, a website must recognize me automatically, at least after my first login, and must not require me to log in each time I connect using the same computer.	16%	26%	20%	30%	7%
14 I believe that in the future most schools and companies will replace their main websites with portals.	By 2002 4%	By 2004 33%	By 2006 6%	By 2008 4%	After 2008 21%

As the reader will see, portals are much more than next-generation websites. They are a new class of Web-based environments, with deep connections to institutional data systems on the one hand, and equally deep connections to the needs and preferences of their end users on the other. A campus portal can affect the recruitment and retention of students, faculty and staff, and can impact—for better or worse—the productivity of all three constituencies.

Portal implementations challenge and change the ways in which colleges and universities perceive themselves and are perceived by their constituents, both inside and outside the walls of the academy. As the case histories presented here demonstrate, a portal is much more than a technological advance; it is a new paradigm for intra-campus interaction and collaboration. The portal is the higher education information environment of the future.

REFERENCES

campusportals.org (2002). *Results of online survey about the definition of Internet portals*. Retrieved April 28, 2002, from: http://134.68.174.102/s1results/.

Jafari, A. & Sheehan, M. (2001). *Defining Internet Portals: Sharpening the Focus*. Poster session presented at the annual meeting of EDUCAUSE, Indianapolis, IN, October.

Katz, R. N. (2000). It's a bird. It's a plane. It's a ...portal. *EDUCAUSE Quarterly,* 23(3), 10-11. Retrieved April 26, 2002, from: http://www.educause.edu/ir/library/pdf/eq/a003/eqm0038.pdf.

O'Reilly, T. (2001, November). SLAC Symposium on the Early Web. *The O'Reilly Network*. Retrieved April 27, 2002, from: http://www.oreillynet.com/cs/user/view/wlg/907.

Strauss, H. (2000). *A Home Page Doth Not a Portal Make*. Paper presented at the Converge 2000 Portal Technology Symposium, San Diego, CA. Retrieved April 27, 2002, from: http://www.princeton.edu/%7Ehoward/slides/portals_files/frame.htm.

Yahoo! Inc. (2001). *The History of Yahoo!—How It All Started...*. Retrieved April 27, 2002, from: http://docs.yahoo.com/info/misc/history.html.

SECTION I

DESIGNING PORTALS: THEORY AND PRACTICE

Chapter II

The ABCs of Designing Campus Portals

Ali Jafari
IUPUI, USA

ABSTRACT

This chapter discusses the fundamental design requirements for building Internet portals, in particular building portals for educational institutions or so-called "campus portals." The focus of the chapter is on understanding portals and their design requirements from both functional and technical perspectives for educational applications. It is meant to offer understanding and to share know-how and experience with those who are involved in various aspects of the design, development or implementation of portal projects.

INTRODUCTION

When the Web was introduced to colleges and universities in the mid-1990s, one of its initial applications was to create campus homepages as gateways to the institution's few and generally disparate websites. Higher education's early websites were very simple to use, but only a limited amount of information was made available on the top-level campus homepage. A campus homepage initially consisted of a nice big picture of the campus or the chief executive and a few links to general brochure-like information. It was designed mainly to provide information for outsiders and for prospective students, and its links were limited to perhaps only a dozen secondary pages. A visitor could explore all the pages of a campus website in less than an hour.

Very soon, however, institutions realized the potential of the homepage as a gateway into the vast information storehouses that universities are. Our homepages became cluttered collections of nested menus linking to hundreds of campus Web pages. This profusion of options made our campus homepages difficult to use, and the typical campus website became an environment unfriendly to users. As a remedy to this, schools began to put search engines on their homepages to help users find their ways to desired information. Quickly, however, the search engine, even one with advanced search features, became useless. Campus websites simply offered too much information and their homepages offered too many links.

The year 2000 witnessed many schools switching to a new design framework for campus websites in order to reduce the usability difficulties. The new scheme categorized information and resources for different groups according to their specific roles and interests. For example, prospective students are a group of visitors who are mainly interested in information such as admission requirements and degree programs; current students, on the other hand, come to the homepage looking for registration information, online library resources and news about what is happening on campus. Many campuses changed their campus homepage designs to include a prominent menu of links to homepages custom tailored for major user-role groupings, including prospective students, current students, faculty, staff and alumni.

In January 2000 and 2001, I conducted surveys by visiting 100 randomly selected university websites. In 2001 found a 15% increase in the number of top-level homepages offering role-based links. Nevertheless, again because information is being added to websites at a near-exponential rate, this role-based homepage design reduced Web usability problems for only a short period of time.

Even as some website designers were trying to redesign their campus homepages with role-based, menu-driven interfaces, a few campuses began exploring a completely new concept, which would come to be known as the Internet portal. The concept was simple but innovative. Using new programming tools such as common gateway interface (CGI) and Active Server Pages (ASPs), campus Web designers developed interactive services linked to their campus back-office database systems. To use a typical portal application of this kind, the user was required to log on. By looking at the user's log on information and comparing it with the information residing in a campus database, the portal immediately identified the user as a member of the campus community, identified the user's role and dynamically presented him or her with a role-based Website that was appropriate and optimized for the user's needs. For instance, users with student status could be directed to a page optimized for student use. Users with faculty or staff status could be directed to different pages, optimized for faculty and staff needs. In a matter of weeks, a programmer could write programs that not only would identify a user and link him or her to an optimized page, it also displayed personal information such as the list of courses that the specific user had registered for, the number of e-mails in his or her e-mail box, the amount of money he or she owed the bursar's office or the parking operation and the types of news that would be most interesting to the user. Portals quickly became the new buzzword on campus and in the higher education information technology environment, became the hot discussion subject in technology publications and at conferences, and created a market for new technologies offered by dozens of vendors.

CAMPUS WEBSITES VS. CAMPUS PORTALS

The differences between a campus website and a campus portal should be viewed from two different perspectives: functional and technical.

Functionally speaking, a campus website offers the same information and same resource to whoever visits it. A website does not care who the user is. In this way, it is like the *New York Times*. Anyone buying the newspaper from any newsstand in any city gets the very same paper with the same front page, pictures and stories. A portal, on the other hand, is designed to recognize a visitor based on his or her role, status and personal preferences. On this basis, it offers different sets of information and resources personalized to the user's anticipated needs. In order to operate in this way, the portal must ask for the user's identification information and requires at least an initial authentication.

Technically speaking, portals offer active and dynamic services as compared to the passive services provided by traditional websites. Databases provide back-end services in all portal environments. The database may hold a single table just to identify a member's role, or may have access to a bank of tables within a large number of databases in order to identify each member and intelligently offer him or her a personalized set of Web services.

SELECTING THE RIGHT PORTAL SOLUTION

Every campus faces the same question before it implements a portal project: what is the best portal solution for the institution? It may take months to find the right answer through the hard work of many individuals.

Following is a table of measurable characteristics of portal systems as set up for three hypothetical proposals. Please note that some of the "Required Characteristics" might not apply equally to every portal project. Position on the list can be used to reflect the priorities of the characteristics. Upon the completion of this analysis, each cell within the table should include a quantitative measure. For instance, numbers from one to ten could be used, if accompanied by some explanatory text.

Selection Characteristics

Ease of Use. I have intentionally placed the ease of use requirement as the top item in my list because I have become convinced of its importance through my many years of experience in the development of information technology (IT) systems for

Table 1. Portals Proposal Evaluation Matrix

Required Characteristics	Portal Option 1 (e.g., proposal from vendor 1)	Portal Option 2 (e.g., proposal from vendor 2)	Portal Option 3 (e.g., homegrown portal project)	Overall Characteristic Analysis
Ease of use				
Maintainability				
Potential for personalization				
Availability of single sign-on authentication				
Ease of customization				
Ease of integration with existing services				
Platform independence				
Performance				
Expandability				
Conformity to open standards				
Availability				
Favorability of pricing and licensing terms				
Viability				
ADA compliance				
Others				
Overall option analysis				Discuss best option

educational applications. First, if a system is not easy to use, it will be ignored by the users. I am sure that many of us in higher education institutions can think of many IT projects that failed mostly because they were not easy to use. Second, the institution will be forced to offer costly training sessions and develop instructional manuals for a hard-to-use system. Third, the institution will be forced to expand help desk resources if its portal system is not easy to learn and use. This could be a noticeable expense because a portal requires a 24/7 availability.

Oncourse was my first enterprise Web application project. It was one of a category of projects that later came to be known as course portals. *Oncourse* is an enterprise course management portal system that dynamically creates a course website for every course and automatically offers access to every student and faculty member at each of the eight campuses of Indiana University. It was developed in the late 1990s in my WebLab at the Indiana University Purdue University Indianapolis (IUPUI) campus. My top three design requirements for *Oncourse* were ease of use, ease of use and ease of use! It was certainly that aspect of the *Oncourse* project that made it spread throughout the IUPUI campus and other campuses of Indiana University much faster than other course management solutions did at other institutions. For instance, in the fall of 1988 when we offered *Oncourse* as a beta environment at the IUPUI campus, more than 600 course accounts were created serving more than 12,000 students. Only 37 faculty members within this group received workshop training the other 200 just figured it out on their own. The ease of use and dynamic enterprise nature of *Oncourse* certainly contributed to its success.

Maintainability. Once a portal system is in production, it requires routine maintenance. The routine maintenance is always being mistaken for hardware maintenance and backup of the database. Routine maintenance includes the maintenance of software and databases that may constitute a major part of a portal's support. The questions are how much maintenance a system needs, how often, with what cost, by whom, the internal staff or external vendors?

Potential for personalization. One of the unique characteristics of Web portals compared to traditional Web home pages is that portals can adapt to the individual characteristics of the user. The more fully the portal adapts, the better. Personalized portals are easier to use, less costly to support and more interesting to visit than static Web pages. A personalized portal automatically offers different information and resources to different users based on their roles, rights, interests, past usage, etc. We are in the very first stages of using the new generation of portals with personalization capabilities. The limits of what we can do with them have yet to be fully explored.

Availability of single-sign-on authentication. A good portal should not ask its users to authenticate more than once. For instance, single-sign-on authentication should let students access information about all of their current courses, register for new courses, access their campus mailboxes, pay their parking fees, use an instant messenger, renew books borrowed from the library and the like. Even though these services may be based on different computer hosts, each requiring authentication, the user should need to authenticate only to the portal and rely on that application's authentication coordination feature to enable transparent access to all the services for which the user is eligible. A portal that understands the user's role can provide automatic authorization, offering appropriate access permission and editing rights and privileges to various services according to the user's right and role within that application. For instance, if I am the instructor of record for the Intelligent Agents course, the portal should automatically offer me sufficient privileges to edit the online contents of my course. Similarly, the portal should let me give grades to my students, look at my students' profiles, etc. Or suppose that a staff member within the campus news department has responsibility for posting news on the campus homepage. Once she authenticates to the portal, the portal should automatically give her a link, somewhere on the portal's personalized main page, that takes her to a data-entry screen she can use to post news text and images on the front page of the campus website.

Ease of customization. Although portals are not the only type of software that offers customization, many believe that customization is only available in a portal environment. Microsoft Office has offered customization for a number of years. Customization is a nice feature offering certain predefined types of software and various settings to different groups (customers) in a portal environment. For instance, the capability of customizing the personal portal page (MyPortal) by moving the e-mail channel to the top left corner, the bookmark channel to the left side, the news channel to the middle column and the weather to the lowest part of the page.

Ease of integration with existing services. Any portal system should have features that allow it to be integrated with existing databases and Web application services. For example, a campus portal would have very little value if it weren't integrated at least with the campus student information system. Before a campus portal project is begun, the design team should identify all the existing databases and services with which the portal is to be integrated.

Most portal projects, when initiated, are broken into different phases, each with a different deployment schedule. Each phase may include the integration of existing services into the portal. For example, the integration to the campus library system may not have sufficient priority to be included in the first phase, but it may be in the plan for inclusion into the second or third phase of the project.

It is important to keep in mind that integration is a two-way street. The campus will want to integrate future services and products with the portal system. The "hooks" a portal product offers to facilitate integration include application programming interfaces (APIs) and adherence to open software development standards.

Platform independence. Many vendors of portal systems tie their products to the operating systems or hardware products of particular manufacturers. Sometimes this is because a partnership exists between the two companies. Other times it simply reflects a technology preference or a business decision on the part of the portal system vendor. Selecting a portal product that is tightly tied to a particular operating system is limiting but may be practical, especially if that operating system is well supported on campus and has a bright future. In most cases such a decision does not also tie the portal to a single hardware vendor. The flexibility to change hardware vendors in response to market conditions is important, so a portal decision that ties the campus to one hardware vendor is not just limiting but dangerous. Unless the campus IT environment is single-vendor focused, and is projected to remain so for the life of the portal product, platform independence is an important selection criterion.

Performance. The performance of the portal is critical when the system is under heavy use. Performance problems become noticeable to the users when the system slows down or rejects new log ons. Performance problems tend to occur during periods of peak use of the system, for instance at the beginning and end of a semester. When implementation decisions are being made, the team needs to consider such factors as network capacity, performance benchmarks of the hardware being considered and the expertise of the system administrators in tuning the performance of the operating system on which the portal application will run.

Expandability. Every portal product analysis should seriously consider the expandability of the system. A portal, at the beginning of its deployment, may not be subject to heavy use. However, any successful portal project is likely to become a "victim of its own success" and to need periodic expansion during the course of its operation. Complex economic considerations will dictate the capacity and performance of the initial portal system deployed, but an intelligent implementation will assume that the system will need to be expanded and upgraded as its users become increasingly reliant upon it, and will select the components of the system accordingly.

Conformity to open standards. When portal system vendors and developers use open programming and interface standards, they dramatically increase the ease with which their systems can be integrated with products created by other vendors. One of the most familiar open programming standards is Open Data Base Connectivity (ODBC). ODBC allows portal developers to write applications and tools that work with any database (e.g., student records, user directory) that supports ODBC. Because of the thriving ODBC development community, every major relational database is now compliant with the ODBC standard.

Availability. Availability refers to the readiness of the technology for actual production use. "Out of the box" readiness is rare, even for mature commercial products. Most of these are highly configurable and require a great deal of pre-production setup. Availability may be an even bigger concern for home-grown portal software. The beta version of such a product might prove itself ready in a limited testing environment but not be ready for use in a production environment. To make sound, informed decisions, implementers need an exact understanding of where their chosen product stands in its life-cycle.

Favorability of pricing and licensing terms. With more and deeper budget cuts facing educational institutions, it is important to have a full understanding of the portal project's implementation and support costs. In most cases the cost of software licensing is the smallest part of the total equation. Day-to-day maintenance of the software and the underlying databases, user support and system upgrades all require considerable human resources and budget. If a campus does not purchase maintenance and upgrade contracts from its portal software vendors, it should seriously consider having on board a strong technical staff and management team. Calculation of the overall budget for implementation and maintenance of a portal project is not easy. Many costs remain hidden until the portal project is in production.

Viability. If a campus is planning to build major part of its portal system using its internal resources, it should pay special attention to the viability of the development group. Dependency on a single developer should be avoided. Because staff turnover is ultimately inevitable, the project manager should insist on complete documentation of the code developed. This may be a bigger concern for smaller campuses with small numbers of IT staff on the development team.

Similar concerns arise when a campus acquires software and services from a new company or one that is facing financial difficulties. A common safeguard in such a situation is to require the vendor to place and maintain in escrow a copy of the source code for its portal application. This won't entirely avert disaster if the company goes out of business, but it will provide the campus with the means to begin a recovery.

ADA compliance. It may not be practical to design every portal application for compliance with the Americans with Disabilities Act (ADA), but by using modern design techniques, the cost of compliance can be reduced. For example, if the portal environment supports features such as cascading style sheets, it can provide a set of templates incorporating fonts that comply with the ADA.

STEPS IN BUILDING A CAMPUS PORTAL

The process of building a portal can be divided into four steps: define, design, develop and deploy.

Define. One of the earliest tasks in a portal project is to define the functional requirements of the system. Requirements may be gathered from different user groups and the leaders of academic and service units associated with the campus. Among these groups are the faculty, students and staff, in addition to outside groups who would use the portal, such as alumni, prospective students, etc. Academic and service unit leaders include department chairs, deans, the provost, library executives, computing services executives, registrar, bursar, admissions executives and the like. The best way to accomplish the task of defining a campus portal is to initiate a committee or task force to include representatives from these core campus groups.

The deliverable of the "define" phase is a conceptual document identifying all major functional requirements for the portal.

Design. As will be discussed in a later section, the design of a campus portal is the most important and critical part of the project. The design should be done by individuals with expertise in technology and an understanding of end users' needs. The design of a portal should follow the functional requirements architected by the "define" group, with input and guidance from the senior software engineer who has been given the role of system architect. Individuals participating in the design phase include engineers and experts in software, database administration, security, user interface technologies and hardware. The design group can be a mixture of internal staff and external vendors and consultants.

The deliverable of the design phase may include comprehensive technical documents and blueprints describing all major technical requirements and procedures necessary for the development of the system and the integration of its components.

The 80/20 rule. Over the last couple of years I have found myself talking a lot about the "80/20 rule." My talk of the 80/20 rule gets longer and longer as I read or hear about portal projects that bypass the design phase and jump into the development and deployment phases of the project. I believe that 80% of the success of a portal project depends on the quality of its design work and the forward thinking put into the conceptual and technical architecture of the system. The other 20% of the project's success is due to the quality of the portal software – the engine that a campus buys or develops using its internal resources. It is not unusual to find a campus portal committee spending most of its resources and time comparing different commercial portal software products rather than committing their time to consideration of what the campus needs, the services it wants its portal to provide and the changes that should be made to the IT infrastructure of the campus to support the implementation of the new portal environment. Campus groups including administrators, service providers, and end users can offer substantial input to the design of a portal project.

Develop. The development phase of a portal project should begin after the project is fully defined and is conceptually and technically designed as elaborated in the previous sections. The development phase of a portal project does not necessarily mean that a campus is "developing" the software rather than buying a commercial system. A campus portal is a project, it is not a product that can be entirely purchased from one commercial vendor. A portal is a system that integrates many different technologies and Web-based services. A campus may buy a large portion of a portal system from a commercial vendor. But these pieces must be integrated with other products and services that a campus has developed internally and would like to continue to develop and maintain. A high quality portal from a commercial vendor can be linked with existing institutional databases through the APIs that the vendor provides.

Should a campus buy the primary components of its portal project or should it build them? There are three obvious options: buy, build and a hybrid of those two. I believe that every portal project eventually takes a hybrid approach, combining commercial components with others that are internally built or customized. The proportions of bought and built components vary widely. Let me say again, however, that I feel it is a serious mistake for campuses to contract out the conceptual design aspect of the project.

Deploy. The deployment of a portal project includes the actual installation of the portal software and delivery of a set of production services. The deployment requires offering services associated with the day-to-day operation of the system, including hardware and software maintenance, help desk support, system upgrades, etc. Some of these services may be contracted out to outside vendors or technology service providers. Many of the services require an ongoing commitment of time and effort from staff in a wide variety of campus offices. This is among the most important reasons for including on the design team representatives from all the offices that will supply information or services to the portal.

THE SYSTEM ARCHITECT: KEY TO A GOOD PORTAL DESIGN

Who should design a campus portal--a system architect, the vendor or a committee? A wrong choice could result in a weak portal framework and system that does not meet the functional and technical requirements specified in the project definition.

System architect. Building a campus portal is very similar to constructing a new building. Therefore, procedures used in a building project can offer ideas and suggest processes for building a campus portal.

Building projects and portal projects both offer various services to campus communities. Both types of project require different types of expertise in their design and development phases. In the case of a building project, we need structural engineers, mechanical engineers, heating and cooling specialists, interior designers, etc. In the case of a campus portal, we need system engineers, software engineers, database designers, interface designers and the like. For both types of project, we buy and integrate some ready-made materials. For instance we buy windows, light fixtures, furniture, etc. for a building and we buy database software, computer hardware, e-mail clients, portal engines, etc., for our portal projects.

One major difference, however, between these two types of projects is the fact that in every building project we DO use an architect but in campus portal projects we DO NOT always assign someone the role of architect. I know of many portal projects in which this role has been assigned to a committee or group of people who lack sufficient and appropriate knowledge and expertise. In other cases, the vendor who supplied the portal software filled the vacuum and played the role of architect for the project.

Who should play the role of an architect for a campus portal project? A campus portal architect should:
- Have a comprehensive understanding of Internet and IT services and their application to the mission of the university (teaching, learning, research and creative activity, and outreach) and the complex administrative needs of the higher education enterprise
- Understand the IT needs of faculty, students and staff
- Be very imaginative and able to foresee the future development of information technology
- Have a good knowledge of various IT technologies and their comparative advantages and potential integration challenges in a portal system.

Like the building architect, the portal architect does not necessary need to be expert in every aspect of IT. A typical building architect has just enough working knowledge of mechanical engineering, structural engineering, interior design, etc., to help him or her make design decisions.

A campus portal architect can be a member of the professional staff, administration or faculty whose experience adequately supports the role. I don't agree with giving this task to a number of individuals or to a committee. A committee may be assigned to offer direction to a portal architect. Such a committee could analyze the different design options suggested by a portal architect. It could review the final design options submitted by the portal architect and make recommendations to campus authorities. But the committee cannot and should not itself do the design of the campus portal. The portal architect should be held responsible for the design and architecture of a campus portal.

INDICATORS OF A GOOD PORTAL

The following indicators can be used to identify a successful portal project.

The portal should know the user. A good portal should recognize all of its members (authorized, active users) and provide resources and services to them based on their role, interests, permission right and preferences; the more a portal knows about its members, the more personal and useful it is. Portals know their members through various sets of data made available to them either manually or automatically.

The portal should be "sticky." A good portal is sticky. It has useful features, offers needed resources, and has a kind of look and feel that encourages users to come back and use it again and again. Portal architects, designers and administrators should always be alert to new services and features that would make a portal sticky, even services and features that may not be directly related to the mission of the project. For instance, the portal could include an eBay-type electronic auction service where students can sell their surplus stuff and look for things offered by other students on their own campus. From a faculty perspective, a good portal could offer services to post notices of campus lectures to lists of scholars whose previous use of the portal suggests that they are interested in the speaker's field of research.

The portal should be well used. A good portal should be regularly used by members of an institution. Low usage of a portal is a clear indication of bad design or poor maintenance. It indicates that the portal is not meeting the functional or technical needs of its users. Poor usage of a portal might also mean that it duplicates a service offered elsewhere on campus.

The portal should offer a gateway into most campus Web services. A good portal offers the campus a single gateway into most or all of the Web-based services and resources a member needs.

USABILITY, A KEY TO PORTAL SUCCESS

I am a big fan of Amazon.com. I like it because it is easy to use, easy to learn, it knows me and understands my needs, and to some extend it is "smart" and shows me only advertisements that I want to see. Many of my colleagues like Amazon.com too, especially those who are not technically oriented and who have limited Web experience and computer knowledge. They like it because it works and they don't need to make phone calls or ask friends to show them how to do things or find things. Amazon.com is among the very few portal sites that I give a grade of A for the design and usability of the environment.

Usability is the extent to which a system supports its users in completing their tasks efficiently, effectively and satisfactorily. Usability may also include an aesthetic component. On the Web, usability extends to factors such as speed, intuitiveness of navigation, clarity, ease of use, personalization and readability.

A growing body of literature addresses the subject of usability for websites and homepages, but very little can be found that focuses on portals. I recommend using www.useit.com, the site of Jakob Nielsen (Nielsen 2002), where the reader can gain a general understanding of Web usability and find guidelines for implementing usability principles. The following section focuses on those aspects of usability that are unique to portals.

Usability requirements of campus portals are more extensive than those for a campus homepage. Campus and commercial homepages are typically designed for public use: anyone from anywhere may visit and should be able to use and easily navigate throughout the environment. Campus portals on the other hand are intended primarily for internal constituencies, members who have usernames and passwords and most likely have extensive knowledge about the institution and the Web services and resources it makes available.

A large proportion of visitors to a campus or commercial website are first-time visitors who can be considered "novice" users. They may or may not return to the website in the future. Users of campus portals, however, are expected to use the portal regularly—as often as several times a day. The majority of visitors to campus portals are expected to quickly become "experienced" users as opposed to the typical novice and casual users of websites and homepages.

Campus portals offer a suite of feature sets that are not available from typical websites or campus homepages. Examples of these features include personalization, customization and the like.

Campus portals use different Web technologies, and use dynamic architecture to provide active environments, as opposed to the much more static, passive nature of the websites and homepages. Campus portals are profile-based environments that require authentication before offering services.

Usability Testing of Internet Portals

Usability testing of campus portals should include evaluation of the unique characteristics of portals as mentioned above, in addition to general measures of Web usability.

The following is a list of variables that may be useful in usability testing of Internet portals. Please note that some of the variables are also commonly used in usability testing of traditional websites and homepages.

Authentication time. How long does it take to authenticate to the portal environment and enter into the top-level portal page (MyPortal)? How much longer does it take to authenticate during the peak times of the day? How long does it take during peak periods of a semester? (Authentication logs can help you estimate the amount of time a single authentication takes.) Can you simulate a worst-case scenario to measure this variable? Are there other services whose performance affects authentication time, for instance, external services like e-mail and news servers furnishing resources into the personal portal pages (MyPortal)?

Single-sign-on authentication. How many of the Web services within a portal environment can be made available on the strength of the initial login authentication? In a typical faculty or student situation, how many times must they authenticate to access a typical set of services (see their current courses, register for next semester, read e-mail, pay bills, etc.)?

Role-based authorization. Does the system offers role-based authorization to various applications and services offered within the portal environment? For instance, do faculty advisors of record get automatic rights to their students' enrollment information, transcripts, etc.?

Page load time. Does the size of the files making up each portal page, including all the images, graphics, icons, background, scripts, etc., adversely affect the time it takes for the page to load? Does it meet the five-second download standard suggested by Web usability experts? If your campus offers distance-learning courses, how long will it take to load portal pages for students connecting through slow modems or taking courses from an overseas country?

Design consistency. How much user-interface design consistency exists among the portal pages? How much of the experience gained by a user in the configuration and use of one application within the portal environment can be applied to others?

Page design. Does the overall interface conform to accepted user interface design requirements.

Single-click shortcut back to top-level page. Does each portal page offer a hyperlink shortcut back to the top-level page? If not, why not? Does the "back" button of the Web browser work as expected throughout the entire environment?

Level of personalization. How automatic is the updating of the role-based features of the portal? If an undergraduate student switches to graduate student status, does he or she automatically get access to resources that are accessible only to graduate students? Are new students added to an instructor's course automatically added to the class roster on the instructor's course website? If an employee's role changes, would the employee automatically receive access to the same Web services that others with that role receive?

Level of customization. To what extent can users customize their portal interfaces? Do you offer an initial, default page design to each user group (student, faculty and staff)? Do you offer default customization at the department level within a given role, for instance to faculty members in the department of engineering?

Browser compatibility. Do all applications and tools available through the portal behave the same in both Netscape and Internet Explorer? How about authentication and authorization? What versions of the Web browser are you supporting? Do you offer an automatic notification warning to those who are not using a supported Web browser?

Platform compatibility. Does your portal run the same on all platforms and with all operating systems?

Configuration robustness. How robust is the system against user errors and improper configurations? For instance, can a user delete major applications that cannot be easily reinstalled? Consider a user who deletes the application that provides access to online course websites? Can he or she easily bring it back? Does the system give the user an automatic warning, with instruction on how to bring the channel back, before allowing it to be deleted?

A portal system with poor usability will require more training and more 24/7 user support. Therefore, there is a negative relationship between usability measures of portals and costs of training and routine user support. The easier a portal system is to use, the lower the cost for training and user support (see Figure 1). This emphasizes the advisability of spending a larger portion of the portal project budget on the design of the user interface, as well as conducting usability testing before and after the deployment of a campus portal.

Figure 1. Portal Usability vs. Training and Help Desk Support

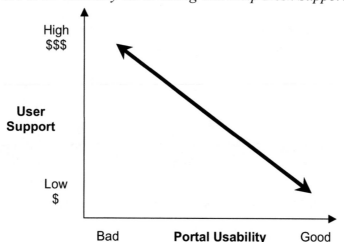

Unlike traditional websites and homepages, portals require a two-stage usability testing process. The first stage should be focused on first-time or novice users; the second stage should focus on experienced users. As discussed at the beginning of this section, portals are intended—and should be designed—to be used very frequently. Users of traditional websites and homepages are likely to remain novice users. Because they will rely on the portal for most of their Web-based information access, users of portals can be expected to attain a higher level of proficiency. A successful portal will facilitate and support this expert level of use.

Usability Testing Methodology

First-time users are the new members of a portal environment who are using the environment for the first time. They should be expected to quickly become advanced users who will use the portal environment many times each day. The best portal environment is one that a new user can sign onto successfully and use easily without needing to read an instruction manual or watch a demonstration.

Laboratory-based testing can be used to evaluate the usability aspects of a campus portal. The laboratory method is appropriate for formally testing usability for both the first-time and advanced users of a campus portal. The following procedure could be used for the laboratory-based testing:

- Identify a series of tasks as elaborated below and create a task sheet.
- Randomly select half a dozen subjects to represent each user group (faculty, student, etc.) and have them participate in the laboratory testing.
- Ask each subject to perform tasks as specified on the task sheet.
- Observe and document the performance of each subject.
- Conduct exit interviews to find out why mistakes were made or why wrong procedures were used.
- Analyze the results.

Obviously, this type of test involves exposing new users of campus portals to a series of tasks and evaluating their success. Each task is a typical use of the system for a member of the test subject's user group. Such tasks might include signing on to the portal, reading e-mail messages, visiting a course site, configuring the news channel, registering for a course and the like. The easiest way to identify and define tasks is to develop a series of scenarios, each representing a typical application of the portal system by a group of users. In each user group there may be subgroups whose use of the portal is somewhat different. For instance, a student may sign on from home, from a computer cluster at school or from a personal wireless laptop in the classroom. A faculty member may use the portal in his office, at home and in the classroom.

Categories of usability problems. Usability problems can be divided into three categories: critical, important and annoying. Critical usability problems are those that prevent a user from completing a task. For instance, it is a critical problem if a user is unable to sign on, access a course or read news. Important problems are those that significantly slow down completion of a major task. Annoying problems delay the user slightly or simply irritate the user without otherwise impeding his or her use of the portal.

Stress testing. One of the important factors in portal usability is the technical performance of the portal under heavy use. This will certainly affect the overall usability of the environment. For instance, if the system cannot handle a heavy usage load, the page load response time may increase or the portal may simply reject new user sign-ons. This is less a concern with websites but is a very serious problem with portals. Using a website generally involves requests for static files of text and a few static images from a Web server. The browser asks for a page, the server sends it to the browser, and the request is fulfilled without extensive processing by the server. A portal page requires much more server processing before a requested page is pushed back to the user. For instance, once a user signs on into her top-level portal page (MyPortal), the portal server must do several database queries and run stored procedures to collect information stored in the user's profile and then custom-build the page before sending it to the user.

Conducting portal stress testing is not an easy task. It can be conducted in two different ways, by a simulation method and by analysis of software code. The simulation method can be accomplished by development of operating system-level scripts that simulate a large number of simultaneous requests to the server. A typical script might request sign-on to the portal, request a personal portal page, retrieve news, etc. The second method of stress testing is more technical and requires analysis of the portal software source code to estimate the amount of server time (CPU time) to complete each task. For instance, it would calculate the number of database queries and the types of queries that would be required to load a personal portal page. This information can be used to estimate the amount of server resources used for every task. Because portal systems are very complex, their performance is affected by many factors, some of which are difficult to foresee in a purely theoretical context. The simulation method, run against a fully operational portal system, will always provide much more useful information than an abstract analysis of the software source code.

Levels of portal usability testing. Portal usability testing can be broken into three major levels. Each level may require its own testing. The levels are:
- *Portal-level usability:* initial authentication, access to various features, configuration of features, editing of profile information, changing of configurations, etc.

- *Site-level usability*: information architecture, site navigation, searching, linking, writing and editing of textual material, etc.
- *Page-level usability:* clarity of the options on the menu bar, consistency of page design, consistency of theme, etc.

An affordable usability exercise. This inexpensive usability exercise should take less than a day to complete. It allows quick identification of major usability problems within a portal project.

Visit a couple of classrooms and ask students what they like and dislike about the campus portal, how many of them use the portal, how often, why they don't use it, etc. Take an assistant with you to tally responses and take notes. I would suggest visiting a large-enrollment freshman class (introductory psychology, history, etc.) and a graduate-level course.

Then ask similar questions of a group of faculty and staff. Visit a couple of academic unit meetings (departmental faculty meetings, schools faculty meetings, etc.) in which faculty members gather to discuss their own issues.

Meet with your campus help desk personnel and ask them about the questions they are being asked most frequently, the ones they find most naïve or uninformed, and who is asking those questions.

If you have a faculty development center or faculty help desk, ask the staff there similar questions.

An ideal team to visit these sites would include a high-ranking IT manager accompanied by a member of the portal team (someone who knows every page and all the features of the portal) and a note-taking assistant. I would suggest repeating this practice regularly, especially a few months after portal implementation and a few months after release of each new version or major enhancement.

Some Observations About Usability

I see many highly used Web applications within higher education institutions whose user interfaces do not incorporate common sense and an understanding of basic user expectations. I am even more surprised when I hear that the institutions responsible for these applications claim to have conducted extensive usability testing. Many of these purportedly tested systems have very basic usability problems that could be identified by someone with even a very basic knowledge of user interface design. For example, I saw a very heavily used e-mail client application within a course management system that does not meet some very basic user expectations for reading and managing e-mails. When I read my e-mail with this system, the interface did not provide an icon to use to delete a message, nor did it show me the date the e-mail was sent. The usability exercise discussed above would have been able to detect these problems.

I still see many popular campus portals that do not offer a direct sign-on box on the front page of the campus portal. Users are required to click on a link or an icon to get to another page before being able to log in to the portal. I don't see any reason why the sign-on box (fields to enter username and password) should not be placed on the front page. On the plus side, putting the sign-on box on the front page of the portal would eliminate one unnecessary click and one unnecessary page load.

DISTINGUISHING BETWEEN A PORTAL AND A HOME PAGE

This might be one of the most controversial issues discussed among portal experts. The majority supports the use of a URL for the portal different from that of the homepage. Very few, including myself, believe the same URL should be used for both campus homepage and campus portal.

My opinion is that the campus portal should take over the well-known and easy-to-guess URL that the campus has been using for its homepage, typically http://www.universityname.edu/. I don't see many benefits to using a different URL to access the campus portal site, such as http://my.university.edu/. Instead, I believe that the campus homepage should function as the top-level sign-on page of the portal by having a portal log on field somewhere on the page campus website. Campus visitors who do not have accounts for the campus portal system can enjoy the generic content available there. Students, faculty, staff, alumni and other groups who do have portal log on IDs can enter that information and get immediate access to the portal. As discussed below, this offers certain advantages and benefits.

One less site to maintain. Instead of maintaining two different websites, one for the home page and one for the portal, which is done on some campuses by two different groups, only one page has to be maintained.

Easy-to-remember domain name. Most people remember or can easily guess a university domain name, and thus can easily find the university's Web homepage. There is no agreed-upon naming convention for the campus portals although many prepend "my." onto their main domain name.

Easier access. Web browser software such as Microsoft Internet Explorer and Netscape Navigator offer only one "home" icon on their navigation bars, not two. Using a single URL for both the campus homepage and the campus portal automatically solves the problem of which URL should be added into a Web browser configuration file as the default homepage for the browser on a given campus.

Automatic exposure to campus activities and news. With the notion that all members wishing to use the campus portal should visit the homepage of their campus in order to log on to the campus portal comes an automatic exposure to the campus website's front page. Today most campuses' homepages include a news and announcements section. By keeping the portal and the homepage accessible at the same URL, an important campus news flash or announcement can be quickly brought to the attention of a majority of campus personnel.

One of the very first universities to implement the single URL convention for both the campus homepage and the campus portal was Brigham Young University. Their multi-purpose website entry page is shown in Figure 2.

Figure 2. Use of Campus Website to Offer Sign-On Access to Campus Portal

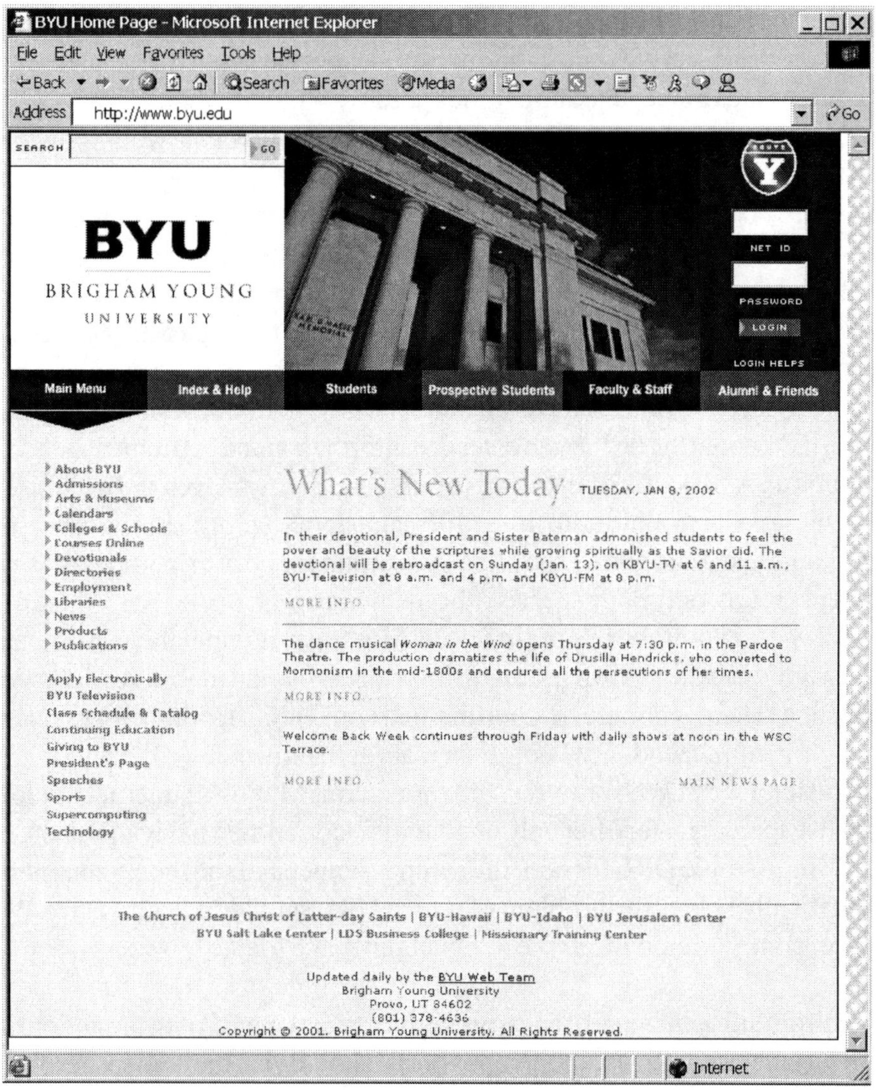

CONCLUSIONS

Portals must include features that optimize users' time in front of the computer. They should encourage frequent visitation by offering the exact services and resources that members require to perform daily teaching and learning activities. Faculty and students spend an increasing amount of time working in front of computers. A good portal design should offer time optimization to every member.

Eighty percent of the success of a portal project depends on the quality of its design work and the forward thinking evidenced in the conceptual and technical architecture of the system—including the human aspects, those things that make a portal sticky, dynamic and help it offer the exact services that its members need. The other 20% of the success of a portal is due to the quality of the portal software or the portal engine that campuses build on their own or buy from an outside vendor. Therefore, it is very important to include the right architect, interface designers and forward-thinking technologists in the design phase of a portal project.

I look forward to a day when every homepage on every campus has a small sign-on box in the top left corner of its front page. After a single sign-on, members receive a portal page that is intelligently personalized and optimized for each individual member of the portal community: student, faculty, staff, parent, alumnus and prospective student.

Chapter III

Keeping Your Eyes on the Prize: Using Inquiry to Increase the Benefits of Institutional Portals

Stephen C. Ehrmann
The Flashlight Program, USA

ABSTRACT

This chapter describes eight potential educational uses of institutional portals (e.g., helping instruction become more spontaneous and adaptive; supporting learning communities; reducing cost of service delivery). It then describes a long-term program of data collection that can improve the educational effectiveness of portals, and control the costs and stresses of portal operation. Studies include: a. baseline data (how well are those goals being met without a portal?), b. debugging studies (what factors are tending to block portal effectiveness?), c. cost studies (what aspects of portal development, operation and use are so costly, time-consuming or stressful that they threaten system success?) and d. outcomes assessment (is portal use contributing to outcome improvement?). The non-profit Flashlight Program has developed a number of evaluation tool kits that can be helpful in doing studies of these kinds.

WHAT IS AN 'INSTITUTIONAL[1] PORTAL'?

For the purposes of this chapter, an "institutional portal" is defined as a tailorable user interface that provides efficient access to an extensive set of institutional resources, communications channels and external resources.

WHY BOTHER TO STUDY YOUR INSTITUTION'S PORTAL?

Without a study no one can really tell whether a portal is educationally successful. So (skeptics might argue) it's safer and cheaper not to do the study and to simply *assert* that your portal is educationally successful. Besides (their argument might continue) if you do a study and find out that your portal has been a waste of money and effort, it might cost your job.

Read this chapter and then decide for yourself whether to do a study. As you'll see, the chapter argues that evaluation can play the same role for a portal that headlights play for a car driving on a twisting road at night: *the right kinds of evaluation can help increase the portal's chances of success and efficiency.*

FIRST STEP TOWARD DESIGNING A STUDY: WHAT KIND OF EDUCATIONAL OR INSTITUTIONAL PROGRESS IS THE PORTAL INTENDED TO ASSIST?

Like a cabinet full of flasks, test tubes and chemicals, a portal can potentially be used for several different educational purposes, depending on choices made by the institution and the users. That will determine the shape of these studies, so we need to define the portal's purpose. Which goals are most important for your institution?

- Enable faculty to offer instruction that is more spontaneous, flexible and adaptive (because they know that all their students are logging on at least once a day).
- Create a foundation for learning communities (by providing effective groupware and providing multiple reasons for people to log on at least once a day).
- Help the users and providers manage an increasingly large and diverse constellation of information for the purposes of teaching, learning and research.

- Save user's time and/or increase their use of services (due to gains in personal efficiency).
- Reduce institutional costs of service delivery by consolidating, reducing or eliminating traditional ways of providing services and using the portal instead (e.g., offering online registration rather than staffing to handle face-to-face registration of all students).
- Help the institution reduce the costs of system change by creating an operating environment that allows systems old and new to interact smoothly with one another.
- Strengthen the bonds with alumni and others outside the community; increase support from these groups for the institution.
- Change student, faculty and staff attitudes toward the institution (the institution is seen as transparent, helpful and supportive rather than opaque and a barrier).

Of course, the portal alone cannot achieve any of these goals. The relation of portal to purpose is somewhat analogous to the relationship of yeast to bread. It's hard to bake bread without yeast, just as it's hard to communicate daily with students if they don't log on, but neither yeast nor portals are the only ingredients in those recipes.[2]

It's tempting to claim "all of the above" as goals for your institutional portal. But remember that actually reaching each of these goals requires a different series of action steps ("ingredients"), and a different set of studies to guide the effort. The more goals your portal seeks to achieve, the greater the expense will be.

The rest of this chapter describes the different kinds of studies that, in combination, can provide a useful and efficient way to guide your institutional portal to functional success. Select those studies that make the most sense for your institution.

STUDIES THAT HELP LAY THE FOUNDATION

If your institution is still considering whether to create (or totally revamp) its portal, it makes sense to find out what other institutions are learning from their experiences with portals. If you can't find a study on this topic, you could do your own. For example you could send an initial set of candidate goals to peer institutions that have had portals for a year or more. Follow up with phone interviews. Ask the respondents to assess the success of their portals in each of those areas. What evidence do they have for citing such a success? (Unless that institution is doing an exceptional job of helping its staff do studies, expect anecdotal information here; it can at least be suggestive, even if it is rarely compelling.) Also ask them about areas of stress and cost during development and operation of their portals.

Studies such as these can help you develop an action plan for your portal project and guide your early work on the other ingredients needed to achieve the highest priority goals. You might learn from this study, for example, that learning communities can be supported and even created with the help of a portal. You might also discover that successful learning communities require many other ingredients, too, some of which may not currently be present at your institution. These might include ways of coordinating student registration in multiple courses, faculty development on how to grade work done by students in teams or creation of new courses. The non-portal ingredients for a learning community, such as those listed above, can take longer to put in place than the creation of a portal. If learning communities are a major reason for creating the portal, it makes sense to begin putting the other ingredients in place as soon as possible so that, as soon as the portal is in operation, it can help create and support learning communities.

The necessity of other ingredients such as faculty development, new online services or new course designs may seem obvious but many colleges have invested in technology and found disappointing results because they followed this route:

1. Some peer institutions bought a new technology (let's call it technology "A"); there was lots of buzz about it. Enthusiasts said Technology A could be used to support learning communities.
2. So this institution bought Tech A, too; discussions of learning communities were then put on the backburner until the system could be made operational.
3. Two years later, after Tech A was deployed and reasonably reliable, discussion returned to learning communities. Someone pointed out that faculty development would be necessary, so, after a few more months, the first small workshops were offered.
4. A year later, other needs had become apparent: new recruitment brochures were drafted, for example, to try to attract students who liked learning communities. Some fixes were needed in space scheduling systems. Change was slow and uneven, however. Money for these investments was in scarce supply. No one had thought to raise such funds, and the new technology had soaked up most of the available funds.
5. Two years later, interest in Technology A had almost disappeared. It seemed slow and outdated. The attention of technology enthusiasts had turned to Technology B, which had 'visualization' as a strength. Learning communities had never really gotten off the ground. Those who noticed this failing tended to blame Technology A which (when compared with Technology B) seemed old-fashioned and weak.[3]

To put this another way, studying what has happened at other institutions can help you define just what the innovation is that you need to plan and evaluate. In this case, the innovation wasn't (just) Technology A; it was the effort to create learning communities, which required Technology A, faculty development, recruitment of students and space planning. Your study of institutions should help you understand the portal-enabled innovations you plan, evaluate and implement.

Such a study should also help you discover what problems other institutions encountered. Your findings can help you avoid some of those problems while preparing users for difficulties that (you discover) are inevitable; people are more likely to endure problems if they have been warned in advance!

BASELINE STUDIES

It is always nice to be able to report that, "we have evidence that our institution is doing (something) much better than it did three years ago," but such statements require that a similar study was done three years earlier: the "before" part of the "before and after" comparison. The "before" picture is called a "baseline study." Baseline studies are ideally done before the portal effort begins, or at least before the portal has had time to begin influencing the outcome of interest. But it's never too late to do a baseline study if gains in the outcome are intended to continue. (Some people may object to the baseline because it's likely to show bad news. But that's the point of taking a "before" picture—to see if the system can help transform 'bad' to 'good,' or 'good' to 'better.')

The baseline study should focus on the behaviors and attitudes that portal availability is intended to influence. That's what determines ultimate benefits and costs of a portal: what students, faculty and staff choose to do with the portal.

For example, if one important benefit is to help instruction become more adaptive and spontaneous (because faculty can communicate with students on a daily basis, for example), how adaptive and spontaneous is instruction before the portal goes into use? How frequently and how effectively do faculty communicate with students before the portal is available?

SYSTEM DEBUGGING OPERATIONS

Debugging studies usually begin early in portal development and operation. They attempt to quickly identify system malfunctions, interface problems, problems in training people to use the system, etc. A system debugging study should investigate potential bugs that would be:

- Important as barriers to one or more of the goals of the portal
- Uncertain (they may happen, or they may not)
- Invisible without a study

EDUCATIONAL DEBUGGING

System debugging refers to failures of system functionality; educational debugging refers to problems in using the portal to accomplish an educational purpose. A portal may appear to work smoothly and yet be found to be buggy when users try to employ it for a specific educational purpose. Such bugs are important to discover because portals only have an educational benefit when they enable actual (not just potential) changes in the nature of educational activities.

For example, one educational goal for an institutional portal might be to help the instruction to become more responsive and adaptive (because the portal has helped insure that students check their websites and mail on a daily basis). If students are discovered not to be using the portal daily, the next step is to investigate potential causes for this educational bug. Likely candidates in this case:

- Not enough important services are easy to use on the portal so some students are not logging on.
- A small number of students are having problems with their Internet service providers, enough students to disrupt faculty plans that depend on quick interaction with *all* students in their courses.
- Some faculty have not yet realized how they could modify the basic structures and strengths of their courses once they begin to use the portal to interact rapidly with students between class meetings.

Some of the bugs discovered may be easy enough to fix. Other cases may be so severe and stubborn that the goal itself must be revisited, redefined or eliminated.

REDUCING COSTS AND STRESSES

Portals are likely to create a shifting pattern of stresses on time and budgets. What's most dangerous about a 'stress bug' is that it can sometimes lay hidden by the enthusiasm of early adapters and the expectation that things will be difficult at first. Studies can provide important early warning. Without a cost study, users may have become exhausted and resentful, and budgets may have been exhausted by the time the problem becomes obvious.

The aim of such studies is to "unstretch" resources: to provide early warning of activities that are demanding disproportionate and unsustainable amounts of money, time or goodwill. You then can use the study's insights to redesign those activities before it's too late.

The typical approach to such studies is called activity-based costing.[4] The study's objective is to gauge all the resources required to carry out a particular activity, no matter which budget and institutional unit those resources come from. For example, a study might focus on costs of online registration for, and dropping of courses. These costs might be distributed among the offices of the registrar, bursar, IT services, student affairs and others.

MONITOR CHANGE IN ACTIVITIES AND, LATER, IN OUTCOMES

This type of study is perhaps the single most important way to improve the benefits of investments in an institutional portal: track and analyze the activities that the portal is intended to enable.

For example, an institution might study whether portal use is contributing to community building. Here are some key activities and outcomes the institution will want to track over time:
- Do users employ portal features to find or work with other people?
- With whom? People they would have worked with before?
- Does use of the portal seem to alter the interaction in ways important to community building? For better? For worse? For example, do the communications seem to help build an appropriate feeling of obligation among those who work together?
- Are there barriers hindering or preventing this type of communication?

One crucial point: even if the portal does help people work and play together in ways that build community, those changes in behavior will probably be apparent months or years before desired community outcomes appear (e.g., increased alumni giving).

For that reason, early studies will focus more on activities (behavior) while later studies will begin to collect more data on outcomes that can then be compared with baseline data.

RESOURCES FOR EVALUATING PORTALS FROM THE FLASHLIGHT PROGRAM

Some of the tools of The Flashlight Program, which I direct, may be helpful in studying portals. Flashlight currently offers several kinds of tools to subscribing institutions including the Flashlight Current Student Inventory (almost 500 validated questions for use in surveying or interviewing students currently enrolled in a course), the Flashlight Faculty Inventory (items for surveying or interviewing faculty) and Flashlight Online (a Web-based system for tapping items such as those to help create surveys which can then be administered either on paper or online). Flashlight Online, for example, could be used to create studies about the portal that could be offered both through the portal and also on paper. Site licenses for the Flashlight Cost Analysis Handbook are also given free to subscribing institutions. Flashlight also works with interested subscribing institutions to help them develop tailored studies; by the time you read this chapter, Flashlight may be working with subscribers to develop study packages for improving institutional portal use. Visit our website at http://www.tltgroup.org.

CLOSING THOUGHT

All too often in the past, an institution bought a technology because the technology is 'in' and enthusiasts demanded it. "We can't compete without it," they might have said. The educational goals (or other institutional goals) for the investment may never have been made clear. And there often was never an evaluation to help the innovation navigate safely through the shoals of implementation. Technical failures are sometimes easy to detect and fix. Educational bugs are often more subtle, and may be experienced by people who don't have the information or budgets to fix the problems. It's difficulties such as these that have sometimes prevented previous innovations from having much impact on institutional teaching and learning. Portals could be an extreme example of this phenomenon. In 2000-2001, I seldom heard clear statements of educational purpose from institutions investing in portals. And I almost never heard of institutions planning to use data to help make sure their investments had an educational payoff. If this discussion has done its work, you should now be able to judge for yourself just how dangerous such self-imposed ignorance might be.

REFERENCES

Ehrmann, S. C. (2002). Viewpoint: Improving the outcomes of higher education: learning from past mistakes. *EDUCAUSE Review,* (January-February), 54-55. This article is also available online at: http://www.tltgroup.org/resources/Visions/Improving_Outcomes.html.

Ehrmann, S. C., Lovrinic, J., & Milam, J. (1999). *The Flashlight Cost Analysis Handbook*. Washington, DC: The TLT Group.

ENDNOTES

[1] I do not ordinarily use the term "campus" portal. I don't think that "campus" should be used as a synonym for institution, for the same reasons that "classroom" is not a synonym for "course": much of the important activity and many of the important resources and are not located far from the physical space of the campus or the class's room.

[2] For more on implementation and evaluation of long-term, technology-enabled educational improvements, see Ehrmann (2002).

[3] Ibid.

[4] For a handbook and cases on how to do activity-based cost models of educational uses of technology, see the *Flashlight Cost Analysis Handbook* (Ehrmann, Lovrinic & Milam, 1999). For information on the *Handbook* and how to obtain it, see http://www.tltgroup.org/programs/flashlight.html

Chapter IV

Portals: Your Institution's Reputation Depends on Them

William H. Graves and Kirsten Hale
Collegis, USA

ABSTRACT

Whether 18 years old and raised on the Internet or an adult seeking the convenience of online service, today's student expects personalizable, online self-service, along with high-touch access to help when self-service falters. Personalizable, online self-service is the promise of the campus portal, a promise that can be achieved and afforded if colleges and universities take seriously the challenge to transform and redesign the form and substance of their high-touch interactions with students and other stakeholders.

INTEGRATED, COMPREHENSIVE, PERSONALIZABLE SELF-SERVICE

"Portal" is a word that should be understood first, not so much as a technology, but as a means to unify three aspects of a quality service environment: 1) the horizontal integration of a comprehensive set of services, 2) the personal customization of those services at the discretion of the service receiver and 3) self-service. Portals enable the integration of services in an online self-service environment and can help improve customer satisfaction through the personal customization of those services. Customer satisfaction derives not only from the flexibility of self-service and the "one-stop shopping" enabled by horizontal integration, but also from the desire for personalized service that takes each individual's service needs and privacy needs into account. However, self-service and personalized service have heretofore been more in opposition than not. On the one hand, "self-service" has typically described a service environment in which there is little or no human mediation—the supermarket in which you fill your own shopping cart and hope not to need help from store personnel until you arrive at the checkout line. (Even the checkout line is now being disintermediated by bar codes and new technologies that read bar codes to total your choices and charge them to your charge card or bank account.) On the other hand, "personalized service" has typically implied a high degree of human mediation—the clerk who recognizes you when you enter the store, knows your preferences and mediates between those preferences and available retail choices, and is genuinely interested in helping you make a purchase that truly meets your needs.

An "enterprise" portal—or institutional portal in the higher education context—is an Internet-enabled service interface that enables the convergence of self-service and personalized service and permits the comprehensive integration of the organization's overall service environment—including desirable services that are external to the organization. Harnessing portal technology to serve an organization's mission most effectively and efficiently requires skipping or moving beyond the mere "bolt-on" stage of technology adoption—a lesson apparently destined to be repeated in the adoption of every major new technology. There is only cosmetic gain in bolting a portal technology onto existing service processes because most of these service processes have been designed in a vertical, departmental paradigm that neglects the customer's desire for an integrated, one-stop service process. For example, dropping or adding a course typically requires visiting at least two different offices to complete, as does the financial-aid process, and many readers will recall standing in at least five different departmental lines to register for five courses. In a portal implementation, there will be no cost efficiencies unless most service processes are redesigned and streamlined to drive out unnecessary expenses while integrating service islands.

In higher education, portals can integrate and personalize online not only the processes associated with administrative and financial services, but also the academic processes associated with learning services—instructional services, library services, tutoring services, advising services and so on. Of course, this does not happen automatically but requires a concerted effort at the institutional level—and sometimes at the inter-institutional level—to redesign and integrate various academic and administrative processes comprising the overall educational process.

Consider, for example, that many institutions today separate residential students from distance students. Whether such distinctions should be reflected in the student information system may be an issue. Even so, any such distinction should be transparent to the student in a portal service environment, once the student is in the system.

Portals should be designed to integrate a baseline of core enterprise services as a service foundation on which each academic program can compete on its academic merits without having to reinvent the enterprise service wheel. In other words, the enterprise portal should be a tool that allows each academic program to focus on its core competencies in its competitive academic market. The enterprise portal will become both a universal service expectation and a point of competitive enterprise differentiation—because few enterprises will have the resolve to undertake extensive horizontal service process redesign across their constituent service units. These themes deserve a place alongside any litany of portal features that might be expected in a higher education environment.

THE BASIS FOR COMPETITIVE ADVANTAGE: PRESTIGE OR REPUTATION?

According to a report by Goldman, Gates and Brewer (2001), excerpted in the *Chronicle of Higher Education,* some colleges and universities compete for and on the basis of prestige, while most should compete for and on the basis of a favorable reputation for customer satisfaction. Reflecting briefly on the role played by technology in each of these competitive strategies from the perspective of today's traditional college-age students will lead to the conclusion that portal services are competitively critical in the context of the residential, undergraduate experience. Couple this conclusion with the long-standing demand for flexibility, convenience and responsiveness in the growing market for adult and employee education, and the conclusion is that portal services are a keystone in *any* competitive strategy today. A portal is a window onto the overall service environment, one that integrates services, personalizes them and presents them to the student and other stakeholders in a default online, self-service format that can provide a competitive edge.

Prestige accrues to institutions that succeed consistently in attracting some combination of an academically distinguished student body, a faculty distinguished by the scope of its research funding and an NCAA-championship-level athletic team in a major sport. This describes only a few institutions, and, in any case, prestige is a risky pursuit that is dependent on the scope of the internal resource base and a host of external competitive comparisons. Only ten institutions, after all, will appear in any top-ten ranking! Yet many institutions pursue prestige, probably because they are for the most part internally governed by faculties and leaders holding doctorates from the hundred or so prestigious or prestige-seeking research institutions.

In contrast, any institution can earn a good reputation by consistently focusing on customer satisfaction throughout its portfolio of services. And portals inject immediate customer satisfaction—the basis for reputation—into the competitive equation affecting long-term prestige. Because the Internet provides flexible, anyplace-anytime human communication and the portal provides personalized, disintermediated service transactions, customer-service expectations among the growing population of Internet-savvy students are rapidly increasing. And few students are more Internet-savvy than those who enroll in prestigious institutions as first-year students. They and their first-year counterparts at less prestigious institutions expect a quality educational experience to include flexible, personalized service options that draw on the convenience of anyplace-anytime communications and self-service service processes and transactions. Today's 18-year olds have arrived at college age with a mouse at their fingertips, ready to use the Internet routinely as a medium for knowing and communicating (see the student commentary in Binns, 2001). For them, the Internet is not a new medium to be questioned or otherwise singled out as external to their natural communication and service environments. Their expectations relate directly to the customer-satisfaction expectations that differentiate a good reputation from a bad one, as described by Goldman, Gates and Brewer (2001). Customer satisfaction across a baseline of integrated, online services—recruiting, admissions, financial aid, academic support and so on—is today's foundation for competitive differentiation via quality academic programs incorporating instructional methodologies that take advantage of the flexibility of anyplace-anytime communication and access to learning resources. Even the most prestigious institutions will have to understand and address this fundamental empowerment of the "customer" or find their prestige under fire from disgruntled external and internal stakeholders—students, alumni, faculty, staff and so on.

Customer demand also accounts for the Internet-enabled flexibility now propagating through instructional and other aspects of the educational services offered to employed adults and their employers seeking to avoid travel costs and other inconveniences of place-constrained or time-constrained educational practices. Institutions that do not respond to the increasing expectation for technology-enabled flexibility in the comprehensive educational process risk damage to their reputations.

ACHIEVING COMPETITIVE ADVANTAGE: A PORTAL THEORY

According to Evans and Wurster (1999), website "navigation is the battlefield on which competitive advantage will be won or lost." Although formulated a few years ago in the context of consumer-oriented e-commerce, this website navigation theory can be transformed into a theory of portal navigation and translated into today's higher education context. Evans and Wurster's three dimensions of navigational advantage translate as follows:

- *Richness*: the depth, breadth and personalized value of the information available to the authenticated portal patron and the capacity for personalized self-service transactions and information views.
- *Reach*: the "pull" of the portal as a marketing tool aimed at bonding the portal patron to the enterprise—the college, university or other educational organization.
- *Affiliation*: whose interest the portal represents from the portal patron's perspective—the organization and its internal constituencies versus the portal patron's as a member of one or more stakeholder groups and as an individual.

This summary portal translation of the work of Evans and Wurster on the role of retail websites aimed at consumers is complex because the frame of reference is the stakeholder portal patron—a shifting point of reference. The portal patron may be any stakeholder, such as a potential student, an enrolled student, an alumnus or alumna, a business partner, a tenured faculty member, an adjunct instructor, a member of the governing board, a staff member and so on. The concept of affiliation embodies an argument that the portal must appear to be designed to serve the particular interests of each such stakeholder group when accessed by a member of that group. The concept of richness captures the idea that each such portal view must be comprehensive, integrated and service-enabled (for transactions). The concept of reach argues against the build-it-and-they-will-come position. "They" may not come to the portal just because it exists. The portal must be marketed externally—for all but the most prestigious institutions—and also internally. And capacity to personalize services and information views should be apparent throughout any consideration of richness, reach and affiliation. The portal must have enough value to each individual stakeholder (personalized richness) to attract continuing use—"stickiness." Few college and university website and portals today satisfy these compelling competitive criteria.

A shared portal can also be a natural focal point for partnering among educational institutions and companies to gain economies of scale in infrastructure and software systems, related 24/7 support expertise, common curriculum development and delivery, common administrative services and common marketing. Such "metacampus" partnerships—so named in Graves (1997)—are typically based on some standardized choices of administrative systems and course management systems. Some state systems and higher education coordinating bodies, already consortial partnerships in their own rights, have created variations on the metacampus to increase access to education and economies of scale. For example, the Kentucky Virtual University is operated by the Kentucky Council on Postsecondary Education, the Tennessee Regents Online Degree Programs by the Tennessee Board of Regents and CalStateTEACH by The California State University Office of the Chancellor. None is accredited in its own right, but all provide access to accredited online programs through some variation on the theme of outsourcing instructors, courses and programs from constituent or partner institutions. All provide extensive anyplace-anytime services, from online student services to online academic programs. All have some form of portal services. And all outsource some combination of infrastructure support, application expertise and support, strategic planning services and project management from Eduprise at considerable economies of scale. The e-Army University is the most ambitious of the metacampus constructs and is made possible by the Army on behalf of its soldiers and its self-interest in an educated and stable armed force—a new form of the GI Bill.

With these discussions about the purpose of portals, their competitive role and a theory that outlines their competitive advantages, some discussion of portal functionalities and practices is in order.

EMBODYING COMPETITIVE ADVANTAGE: THE PORTAL OF TODAY

Portals today come in a variety of shapes and sizes. There are vertical and horizontal portals, commercial and homegrown portals, and enterprise information portals (Looney & Lyman, 2000; Eisler, 2001). Vertical portals are large bodies of information specific to a particular topic. Portals specializing in a particular type of poetry, author and/or more broadly a field of study are an example of a vertical portal. Horizontal portals—such as MyYahoo! and AOL—integrate the specialized functions of vertical portals and data stores. According to Gleason (2001), the most powerful portal is the enterprise information portal that "provides a single, intuitive and personalized gateway to access and to integrate campus-specific information and applications with unstructured data from on and off campus." The unification of vertical campus data and the incorporation of external resources define the campus portal that can set some higher education institutions apart from their peers. Whether portals are vertical, horizontal or enterprise, they can be developed by commercial vendors or institutions, or they can be a combined effort incorporating commercial products into a homegrown solution.

Higher education, rightly so, is focusing first on the needs of currently enrolled students. Eisler (2000) cites three main purposes of a portal: 1) to act as a gateway to information, 2) to serve as a point of access for constituent groups and 3) to serve as a community/learning hub. The gateway is perhaps the most significant aspect because it is this purpose that requires disparate systems to be unified beneath the portal umbrella. Although students, faculty and staff often use portals, the features available to each audience typically vary. Most portals allow all audiences access to group-specific materials, such as campus or departmental news or club calendars. Portals also typically offer community-building communication tools. However, the campus portals of today, as gateways of information, do not treat all audiences equally. Currently, most portals have been designed to focus on the service integration needs of enrolled students. Few of the portals on the market today also take into account the needs of faculty, staff, alumni, prospective students and other constituent groups.

Portal vendors and homegrown portal solutions offer authenticated users a space they personalize to meet their needs within an enterprise information context. This space is secure and typically offers single authentication into a variety of campus service systems. These online services commonly are campus news, calendaring, search engines and community-building tools such as discussion boards, chat tools and e-mail access. Authenticated users can customize their portals to add "channels" for their club activities, stock listings, and local and national weather and news reports. In addition, users can turn to the portal to check their e-mail, chat with other institutional constituency groups or collaborate with their peers in campus-sponsored clubs or classes.

To go beyond these basic services, commercial student information system vendors are working with portal vendors to integrate Web-based access for typical student service functions. These partnerships, such as those between Datatel and Timecruiser and between SCT and Campus Pipeline, allow students to enter the portal and register for classes, check their grades, update personal information, review their transcripts and transact many other Web-enabled student services. In addition, Blackboard offers a portal that is readily integrated with its course management system and a host of other systems via open application programming interfaces that allow campus information technology staff to integrate course management systems, student information systems, e-commerce systems and other campus data systems into their portals.

The commercial portal vendors are not the only ones creating these online services. A number of campus-based initiatives have duplicated these services by integrating their homegrown portals with their student information systems. Here are some examples.

- MyUW, created by the University of Washington (http://myuw.washington.edu), integrates a number of campus systems via the portal, including the student information system, housing and food services, the alumni information system, and a system that compiles daily news and weather information. Students are able to log in and check their course schedules, see personal calendars and check the balances on their student food services cards (a.k.a. the "Husky Card"). Faculty members can log in and see teaching-related information including updated class rolls and schedules. All University of Washington employees can use the portal to update their personal data such as address, phone number and emergency contact information. Alumni can log in to see information about their degree programs as well as their current personal data.
- The University at Buffalo, State University of New York, created a homegrown portal solution (MyUB) in 1998 (http://www.buffalo.edu/aboutmyub). MyUB offers undergraduate and graduate students access to almost every student service. Students can register for courses; apply for financial aid; and view class schedules, exam schedules, grades, upcoming university events, and local and national news.
- At the University of Minnesota, My One Stop (http://onestop.umn.edu) was created to offer students, faculty and staff members a personalized Web experience. Students use the portal to access general campus information, academic information and local news items. Perhaps the most advanced feature of My One Stop is the focus on staff members as a portal audience as well. My One Stop offers employees a Human Resources page where they can "display…up-to-date Vacation and Leave Accrual, Retirement Account and Flex Spending Account data."

North Shore Community College (http://pipeline.nscc.mass.edu/cp/home/loginf) provides an example of how an institution can develop a customized solution that extends a commercial portal. North Shore has selected Campus Pipeline as its commercial portal and has worked with its information technology management service provider, Collegis, to integrate the portal with a number of existing campus systems in a custom development environment. Students can perform many of the student service transactions that they typically would have stood in line for hours to complete. They can check their calendars, read and send e-mail, and access online class materials. North Shore has not only focused on the needs of currently enrolled students, but has created a portal environment that can serve faculty members as well. In addition to personalizing their instance of the portal, faculty members can access real-time class data from the student information system via the portal interface. This means that instructors no longer have to wait for published class rolls at the start of the semester. Instead, they can quickly access up-to-the-minute roster reports and verify the students who are registered for their courses.

In addition to the commercial vendors and homegrown portal solutions, there are consortial efforts, such as JA-SIG's uPortal, that offer alternatives to a strictly buy-versus-build decision. uPortal is, according to Gleason (2001), "a framework, a set of technical specifications and software...that will permit individual institutions to customize the institutional portal by plugging in components in a well-defined and usable manner." uPortal offers portal builders a head start through the use of an existing framework. This framework has similar features to other commercial or campus-based portals, including: secure access, gateway services to information, single login to a number of campus service systems, communication tools and provisions for users to personalize the interface.

Today's most advanced enterprise information portals allow prospective students to apply for admissions, check the status of their applications and chat with academic advisors. Current students can perform almost every function currently offered by the campus registrar—including enrolling in classes; reviewing and/or ordering transcripts; conducting degree audits; and paying tuition, library fines and parking fees. In addition, these advanced portals allow faculty to go beyond the basic offerings of a typical portal by accessing class rolls, submitting final grades and looking up student directory information. Faculty and staff may also be able to use the portal to reserve meeting rooms, schedule multimedia equipment, submit purchase orders or travel reimbursement forms, or check on the status of orders and payments. These broad enterprise information portals are the exception, rather than the rule in today's portal environment.

Portal adoption is proceeding apace in higher education. Many institutions are phasing in a portal starting with a horizontal portal focused on enrolled students and growing toward an enterprise information portal that serves all constituents. This is an effective strategy. The challenges of portal adoption and diffusion go far beyond the challenges of technology. Passing disparate vertical data from one system to another is often far easier than engaging the fundamental process redesign agenda necessary to generate the financial benefits that can accompany portal adoption.

IMPLEMENTING THE COMPETITIVE ADVANTAGE: THE PORTAL CHALLENGE

In the age of ubiquitous information, the organization, presentation and maintenance of proliferating data are increasingly more complex tasks. Touted as the panacea for information overload and system incompatibilities, portals are a hot topic for academic organizations. Enterprise information portals are seen as the cure for data redundancy, the way to integrate multiple systems into one seamless whole and the way to provide end users with a personalized, one-stop shop for all resource needs. Although this vision of an enterprise portal is excellent, it is often tempered with a much different reality.

Implementing an enterprise portal effectively will engender a fundamental shift in the way an organization provides services. A portal changes users' expectations for interacting with the organization's systems and raises the bar for the online service environment and the scalability of the infrastructure environment. Planning for the integration of multiple systems and the simplification of data stores is often the focus of a portal implementation, but understanding the magnitude of the portal implementation as it impacts day-to-day service processes is a reality that often comes far too late. Jim Dolgonas, Deputy to the Associate Vice President and Director of Information Systems and Computing at the University of California, was quoted in a recent article (Sistek-Chandler, 2000) as saying, "the most difficult challenge is facilitating a culture change across the institution. Many departments and universities as a whole are resistant to change."

The College of the Holy Cross also recognized this challenge in undertaking a portal planning project and suggested not only that the portal implementation be a phased project, but that the portal be a university-wide adoption in which the college should expect all departments to adopt and adhere to the architecture and vision for the portal. In addition, Holy Cross (Paadre & King, 2001) realized the need to "concentrate on staff training in these new technologies and create a plan to systematically introduce components that are relevant to the College's constituency."

Effective enterprise information portals will fundamentally change how organizations function. With access to more information that is better organized, organizations will be expected to provide additional services to their constituents. For the academic advisor, for example, this means being able to help students determine which classes to take and also being able to help them clear financial stops, renew library books or transact any other service offered via the portal. For members of the faculty and staff more generally, portals offer new and convenient ways to order supplies, reserve resources, collaboratively work online, train for additional skills, and organize their work and professional lives. The organization therefore should prepare all internal stakeholders for their changing roles and help them learn to manage the data that flows through their individualized portals. Processes and roles must be redefined, if not before the implementation of a portal, then in parallel with the implementation.

ENVISIONING COMPETITIVE ADVANTAGE: THE PORTAL OF TOMORROW

According to Gerry McCartney, associate dean and chief information officer at the Wharton School at the University of Pennsylvania, quoted in a recent article (Norman, 2000), "A portal is a place that draws people to it because of what it offers and what it enables." No longer is the definition of a portal limited to the "companies that served as entry points to, and then aggregators and organizers of, the vast realm of resources available through a still nascent World Wide Web," in keeping with Norman (2000). Today's portal offers information specific to a topic or to an individual. Tomorrow's portal will offer even more.

It is not enough to draw people to a portal. To elevate the reputation of an institution, the portal must have "stickiness"—richness and reach that causes users to return again and again. The portals of tomorrow will be cradle-to-grave information founts. Imagine a portal that learns more about you each time you visit. Every search query is stored, and eventually information relevant to your interests, your field and even your family can be pushed to you from your personal portal—in much the way that Amazon.com "knows" its customers and pushes information to them about new products relevant to their interests.

Prospective students will visit the campus via the portal, create a personal profile, and have the information customized to their academic and social interests. The prospect will be able to setup a videoconference or chat session with an academic advisor, faculty member or current student to learn more about the institution. The institution will learn about the student as well by tracking what information is requested through browsing and search queries. The prospect will be able to complete an application online, and be notified via e-mail when it is received, reviewed and when a response is available via the portal. Prior to arriving at campus, prospects will be able to tour buildings, see their dorm rooms, apply for student loans, sign up for classes, order books and supplies, pay for the semester, establish their meal plans, order sporting event tickets and sign up for clubs—all from the institution's portal. Shortly thereafter, they will receive their residence hall keys and their all-in-one identification card—created using a scanned photo submitted and now recorded as part of their personal profiles. In addition, they are e-mailed the profile of their new roommate and invited to a chat session to meet their roommate for the first time. All of this happens *before* they arrive on campus, and most of it is automated.

By establishing the campus portal as the personalized source for all information about the institution prior to admittance, institutions will show prospects more relevant information, demonstrate the institution's commitment to making the student's learning experience a personalized and technologically advanced experience, and establish the portal as the primary access point when that prospect becomes a student.

Current students will use the portal not only for academic purposes—to register, check their transcripts, apply for advanced degree programs, pay for tuition and books—but for personal purposes as well. Today's portals already allow students to add personal channels to check stock information, local weather and news events. The portal of tomorrow will go far beyond that. It is not difficult to envision a time when students will be able to add online banking channels, travel channels, and channels in which they can play collaborative games, operate their personal jukeboxes or view the latest DVDs. The system can actively engage the students by sending out e-mail alerts for key academic events such as registration, tuition deadlines and advising opportunities. It can remind them through e-mail or portal alerts of upcoming homework deadlines and exam schedules, and it can let them know when submissions have been graded and are ready for their review.

Going beyond campus-based information, the portal can be used to engage students with the institution by pushing interactive information to them. This use of push technology can be personalized as well, basing the information pushed on the profile stored in the institution's numerous data stores. The university athletic association and student union can notify students of upcoming events that might be of interest. Students can also receive e-mail before holidays with airline ticket price information for flights home. Students can then purchase those tickets through the portal and have their travel schedules automatically added to their portal calendars. The campus portal can become the student's primary point of entry to the World Wide Web for students.

The transition from student to alumni need not eliminate the value of the portal as the gateway to Web information. Campuses can make the campus portal robust enough to carry students beyond the time of their college-age enrollment. Attracting alums to life-long learning opportunities offered by their alma maters is becoming as important as initially attracting them to undergraduate or graduate programs. A campus can become the student's preferred education provider for a lifetime.

Through the portal, students will use "MyLibrary" features to customize their information resources. The longer they are in school and the more they use the search capabilities available, the better the system will understand their interests. When the library receives new materials that fit a student's personal profile, the portal will e-mail a notification and ask if the book should be held for checkout. The library will also recommend other sources of information, based not just on a single query conducted by a student, but based upon years of that student's search for and selection of relevant materials. Students will be able to rate resources as they apply to their research and interests. And the more a student provides such ratings, the better the system becomes at recommending the right information to the student.

This service can be considered a student's personal, electronic librarian. However, unlike many librarians, the system not only knows the resources available to the student, but knows the student as well.

Once students graduate and leave the physical campus behind (if they attended an institution with a physical campus), they can take the services of their personal, electronic librarian with them into their careers. The portal will have accumulated not only four or more years' worth of data about a student, but will know when the student graduated with what major and into what career or job, provided the student maintains her profile via the portal. As the alums advance in their careers, they will return again and again to the portal.

The preceding example illustrates some aspects of the cradle-to-grave interactions between students and the institutional enterprise portal. However, enterprise information portals will serve all constituents of an institution: prospective students, students, alumni, faculty, staff, adjuncts, administrators, board members, community members, parents and so on. Each of these audiences must be considered when institutions intent on distinguishing themselves through the use of a portal begin their planning.

IN SUMMARY

Campus "enterprise" portals enable integrated, comprehensive, personalizable self-service. Self-service is not the demise of human mediation, but is instead a challenge to rethink the form and substance of human mediation in every aspect of the educational process—typically thereby signaling the need for redesigning a set of vertical academic and administrative services into a unified service process. Accordingly, every educational organization should engage these issues and develop strategies and plans for a successful and ongoing process redesign and portal implementation process. The need for integrated, comprehensive, personalizable online self-service is most obvious for working adults and others who are enrolled in online academic programs and do not have the safety net of the campus classroom and service office, but for the sake of contextual familiarity the emphasis here has been on the "traditional" student. In any case, there is no reason not to provide the highest quality of service to *all* students when services have been redesigned to take advantage of the cost economies inherent in Internet technologies.

As organizations implement basic portal functions by integrating back office systems, course management systems and a few other systems into a single interface, they should look to the future to understand how far-reaching their portals might become and plan for that future today. An organization's enterprise portal may someday not be limited to internal information and transactions, but may have to support the storage, retrieval and manipulation of much of the data associated with a constituent's life. If the locus of the portal's design shifts from the organization and its employees to their "customers," as recommended in Evans and Wurster (1999), then the portal will confer an even greater competitive advantage than now imagined. The portal could then become a life-long bond between the campus and all of its stakeholders—a bond that grows with the individual's web of contacts and intellectual interests to become an ever more integral component of everyday life.

REFERENCES

Binns, R. (2001). Were they ready for me? *Converge*, 4(10), 50-52.

Eisler, D. (2000). The portal's progress: A gateway for access, information, and learning communities. *Syllabus*, 14(2), 12-18.

Eisler, D. L. (2001). Campus portals: Are we ready? Workshop presented at *Connections 2001*, Centre of Curriculum, Transfer & Technology, Whistler, British Columbia, Canada.

Evans, P., & Wurster, T. S. (1999). Getting real about virtual commerce. *Harvard Business Review*, 77(6), 84-94.

Gleason, B. W. (2001). uPortal: A common portal reference framework. *Syllabus Magazine*, 14(12). Retrieved April 28, 2002, from http://www.syllabus.com/syllabusmagazine/article.asp?id=4136.

Goldman, C. A., Gates, S. M. & Brewer, D. J. (2001). Prestige or reputation: Which is a sound investment? *Chronicle of Higher Education*, Oct. 5, 2001, B13-15.

Graves, W. (1997). Free trade in higher education: The meta university. *Journal of Asynchronous Learning Networks* 1(1), 97-108.

Looney, M. & Lyman, P. (2000). Portals in higher education: What are they, and what is their potential? *EDUCAUSE Review*, 35(4), 28-36.

Norman, M. M. (2000). Portal technology: Into the looking glass. *Converge Magazine* (Suppl.). Retrieved April 26, 2002, from: http://www.convergemag.com/SpecialPubs/Portal/portal.shtm.

Paadre, H. & King, S. (2001) *College of the Holy Cross: Electronic Community and Portals*. White paper retrieved November 29, 2001 from http://www.mis2.udel.edu/ja-sig/holycross.doc.

Sistek-Chandler, C. (2000). Portals: Creating lifelong campus citizens. *Converge Magazine,* October (Suppl.). Retrieved April 26, 2002, from: http://www.convergemag.com/SpecialPubs/CampusCitizens/defining.shtm.

Chapter V

Developing a Portal Channel Strategy

Jameson Watkins
University of Kansas Medical Center, USA

ABSTRACT

This chapter provides a method for organizing a portal channel development strategy by identifying potential content, classifying it and then prioritizing it into distinct categories. Several effective ways of identifying content are discussed that include committees, focus groups and pilot projects. Representatives of the campus communities that will be using the portal are important to poll, ensuring they describe their actual needs versus what they think they need. External resources aggregated into the portal must be appropriate to the institution and reliable. Channels that streamline your institution's business processes will be the most valuable parts of your portal; the bulk of your portal development work should be spent in creating them. Understanding your portal vendor's programming interfaces to create custom, integrated applications is vital, as well as their philosophy in distributing new portal channels.

INTRODUCTION

Using Yahoo!, a person can search the Internet, send e-mail, manage finances, join an interest group, create a Web page, track personal calendars and chat with friends. At least half a dozen other sites offer similar functionality, and all are continuously adding new services in an effort to outdo each other. At any given higher education institution, it's likely a large number of people have accounts with and actively use such a commercial portal.

On the flip side, many campus communities aren't using consumer portal services at all—asking a non-traditional student who was nervous about a home computer requirement to use online, self-service applications can be a ridiculous proposal. For even a daily user of the Internet, it's a large cognitive leap to log in and start customizing a portal from the static homepages to which he or she may be accustomed.

From advanced users who have a myriad of choices available to them, to the first time user of the Internet, what would motivate someone to use an *institutional* portal? One of the main factors in attracting users is available content that can't be obtained from another source, or at least not as easily. A university portal's success and wide adoption hinges on a blend of useful services that enable users to organize information pertinent to their roles at the university and accomplish daily tasks.

BACKGROUND AND DEFINITIONS

In the higher education context, *institutional information portals* are applications that integrate campus-specific information and applications with other sources of information from on and off campus and provide a single, intuitive and personalized gateway through which to access it (Gleason 2001). An institutional portal must fulfill the informational needs of students, staff, faculty, alumni and visitors, as well as potential employees and students. Users of the portal may also include staff from other institutions who require access to certain services, or communities of users that the university serves through grant and community relationships.

It's easy to get caught up in trying to redefine the way people use the Internet. Certainly lines blur between the operating system, the Internet and a portal, and the methods in which users interact with them—cell phones, pagers, PDAs and even instant messaging. The point of an institutional portal should not be to take over a user's Internet experience and provide a wrapper or gateway for all Internet content, or, to use an industry term, to "attract eyeballs." Too often university developers get caught in the commercial doctrine of making sites that are "sticky" and entertaining, equating large numbers of page views with success. Users of an institutional portal should come to the site because it is the most convenient way of organizing institutional information and services; rich, valuable content should be the star of a portal, and the measure of success in a portal implementation should be the convenience it offers its users.

One of the identifying traits of a portal is the compartmentalizing of various services within an overall page structure. A common characteristic of many portals is the ability to collapse, expand, delete or change the layout of each service. Naming conventions for these compartmentalized services offered through a Web portal vary widely depending on the portal vendor. Some examples include uPortal's channels, Oracle's portlets, Novell's gadgets, PeopleSoft's pagelets and Microsoft's Web Parts. For simplicity's sake, I will use the term 'channel' throughout the chapter to refer to discreet portal services.

OVERVIEW

Determining portal content can actually begin at several points in the development cycle, and like many other complicated, multi-tiered projects, it can occur simultaneous to other tasks associated with implementation. A typical portal project contains these basic steps:

1. Determine portal requirements
2. Evaluate/demo portal products
3. Select and purchase portal and additional hardware/software requirements
4. Identify and prioritize potential portal content
5. Train staff in development environment and/or hire consultants
6. Develop portal framework
7. Develop portal content
8. Launch
9. Evaluate
10. Refine

The need to publish certain content can often drive an institution to investigate portal solutions, and specific types of content may even drive the adoption of a particular portal vendor. If you've been given the mandate to Web-enable a financial system by the end of the year, you're likely to look hard at your financial system vendor's portal product for quick results. It's important to keep in mind other factors that include cost, time and available skills, however – quick results may be good for completing your short-term problems, but may lead to choosing a product that's not suited to the institution's long-term strategic goals. Absent clear mandates from the executive level, rounding out a full channel development strategy, will most likely occur after a university decides on a portal vendor and the development environment is established.

Channel strategies can be broken down into three basic categories: identifying external information sources to be incorporated into the portal; leveraging pre-built channels created by the portal vendor or other portal developers using the same platform; and creating custom channels to meet business needs in individual organizations. Each of the three categories requires different types of planning and skills to accomplish. There are, though, several methods you can employ to determine what channels you need to focus your attention on that span across all three categories.

IDENTIFYING CHANNELS

Identifying potential content for your portal isn't something that can be done in a vacuum. In this regard, planning for a portal is similar to the kind of planning and coordination involved in an institution's website. It should not be a completely foreign concept to engage elements of the university in determining what features should be available and how they should be presented. Several ways of identifying content on campus include forming committees, holding focus groups, employing a business analyst or creating a pilot portal. These strategies are not mutually exclusive; if you have the time, a well-rounded approach would employ several or all of these strategies.

Committees

To most universities, forming committees charged with making decisions that affect the entire campus is a time-honored tradition and standard business practice. Committees are groups of people organized to address a single goal and have clear starting and stopping dates. The goal of a committee is usually to produce a decision, often in the form of a white paper document that introduces the goal, defines the methodology and, of course, comes to some conclusion. Committees tend to serve several purposes. If chosen correctly, committee members are representatives of various populations that will be affected by the outcome and can reasonably be expected to speak for their constituencies. A second purpose is to gain buy-in from campus stakeholders on the outcome of the decision. If an organizational unit on a campus has a representative at the table, it is less likely to question the outcome of the decision, even if the outcome isn't to their benefit—it's important that they had a voice and their point of view was taken into account.

Committees can be used to determine portal content if handled effectively. Key stakeholders in the portal project should be present—those who will be doing the development work, executive members responsible for funding the project, managers of systems you intend on collaborating with and heads of departments that will be using your portal.

Obviously, there are downsides to decision-making by committee. One common criticism leveled at committees is the pace at which they move—what seems like a relatively straightforward objective can drag into months of meetings. Issues faced with forming a committee can include:

- Scheduling a group of active campus participants to meet regularly can be nearly impossible. Those who miss meetings can bring down the productivity of a group by requiring rehashing of issues that have been resolved previously.
- If not carefully focused, members may approach the problem at inappropriate levels of detail that can lead to hours of debate over minutiae like the wording in a document or the color of a website background.
- Those chosen to serve on a committee may not be vested in the outcome of the decision; a busy stakeholder may delegate committee work to a subordinate who doesn't share the same interest or simply doesn't have the institutional perspective to understand the project.

Take time to identify the membership of the committee. Make sure they understand the importance of the portal and will take their duties seriously and actually have the time necessary to devote to it. Establish the goal of the committee early on and set beginning and ending dates for the commitments you're asking of the members.

Focus Groups

Convening a focus group is a somewhat informal technique that can help you assess user needs and feelings both before content decisions are made and long after implementation. In a focus group, you bring together from six to nine users to discuss issues and concerns about the features. The group typically lasts about two hours and is run by a moderator who maintains the group's focus (Neilson, 1997).

There are several ways of using focus groups. One is to not even attempt to describe what a portal is and how it will operate. Instead draw their attention to the abstract problems portals are trying to solve. In this way, you aren't presenting them with the solution before knowing their problem. This is accomplished by asking general questions pertaining to workflow and how users spend a majority of their time during the day. What tasks are time-consuming for them? What routine pieces of information do they use in a typical day and how do they access them? If the members are in service positions, what information requests do people come to them about on daily basis? Ease into the portal solution from there: would making this information available to individuals in an electronic format solve the problem?

Focus groups have their pitfalls as well. As with any method based on asking users what they need—instead of measuring or observing how they actually work-focus groups can produce inaccurate data because users may think they want one thing when they really need another. To minimize this problem, a completely different tack may be to expose users to the most concrete examples of portal technology possible. This includes providing demonstration portal applications at a workstation in front of each member of the focus group or, if not practical, providing a projected live image of a portal in the room. Walk through the various aspects of a portal and make sure the group members understand its capabilities. Using this method, your focus group may be able to provide you with solid ideas on what would work in a portal environment and what wouldn't.

In forming a focus group, some of the same rules apply as for committees. Attempt to get a good cross-section of your campus community or you'll wind up making assumptions about what users need based on a minority of opinions. If possible, conduct several focus groups with the various populations. Try to bring together a group of students without faculty or administration present—the presence of authority figures may control the flow of the conversation and undermine the expression of subordinate members' true needs.

Business Analysis

A more methodical approach is to single out business processes and analyze them in the hope that they may prove useful to incorporate into a portal. A process analysis could be as simple as counting the number of logins or clicks a user makes to access a particular Web resource, or as complex as detailing who inputs what into an enterprise financial system, what types of reports are run and who uses them for what purpose. Typically the result of such an analysis is a workflow diagram that specifies which tasks need to be executed in what order. Ideally it details at what points in a process additional inputs are required, and where the outputs to a process occur. Understanding a process at this level is critical to making a decision to include it in your portal, especially if it is complex and will take considerable resources to implement.

Business analysis of this sort is particularly useful in attempting to convert into a Web or portal environment a business process that is undocumented and is now done via manual inputs and outputs such as phone calls and paper-based documents. Some sort of analysis will be required of any complex process that is to be converted into electronic media, and the result of an analysis can tell you if a particular process is a good portal candidate. The results are always interesting and valuable if done correctly, but the outcome may be that a particular process should not be integrated into a portal environment. As disappointing as that may be, knowing what won't work in a portal is still useful.

What processes should you analyze to begin with? This is an even larger question than whether a particular business process would work well for the portal. In-depth analysis is a time-consuming and specialized skill, and even the smallest institution can have hundreds of such undocumented and potentially useful processes. The best place to start identifying such processes is in a committee or focus group.

Pilot Portal

Forming committees, moderating focus group sessions and working with business analysts can take a considerable effort, drifting into timeframes of many months, before a single line of code has been produced. Foregoing a systematic approach to planning may actually make the most strategic sense if you're on a tight schedule to complete the portal or need to demonstrate its capabilities to an executive-level office before receiving a commitment (possibly before you get the go-ahead to form committees or focus groups or perform business analysis). Getting a working release out the door and in the hands of users may be the most efficient way of generating ideas for what to include in the final portal project and in gaining important grassroots support for it.

Start with a set of basic portal services that include authentication, bookmarks, announcements, search services and a user feedback mechanism. Include an external news channel and a port of a popular Web service from your existing website. Focus the majority of your development time on one currently unavailable feature that you know from experience will be a welcome and popular service that is not currently available, such as personalized financial aid information. By producing a sampling of what you can do with a portal, you learn an enormous amount about the technology, provide a service that wasn't there before and, most importantly, give your users a working example to stimulate feedback about what else would be useful to them.

Pilots need not be fully functioning, but should represent the goals you're trying to accomplish with a portal and bear a reasonable likeness to your envisioned production version. A danger in providing a pilot portal is that it sets sometimes-rigid expectations of the final product. If the graphics in your pilot are of decidedly low quality, for example, it may not be apparent to those unfamiliar with iterative design that the look-and-feel can change radically without affecting the underlying functionality. One nightmare outcome from a pilot portal that had poorly designed navigation elements or numerous server errors would be users who gave up in frustration, turning potent grassroots enthusiasts into critics.

TYPES OF CHANNELS

Portals can include three basic types of channels: external, pre-built and custom. Each should have a separate development strategy. Because each requires a different skill set, breaking out your resources based on the different types of channels you have chosen may be productive and relatively straightforward.

External Channels

Defining external news sources may provide the easiest and quickest 'gee-whiz factor' you can hope for. Most portals allow administrators to define external feeds of information that are based on a standard format. External channels are an important piece of the portal experience as they demonstrate the portal's powerful personalization capabilities. Though such channels can be seen as extraneous to your core goals, they can still be valuable—if your users are going to several news sources daily anyway, it makes sense to provide them with that information in an efficient manner right alongside services that enable their everyday work tasks.

One popular method of incorporating external channels is by using a format called Rich Site Summary, or RSS. RSS is a lightweight XML vocabulary for describing metadata about websites and is ideal for news syndication. Originally developed to populate Netscape's My Netscape portal, RSS has taken on a life of its own and has become perhaps the most popular XML format today. Thousands of websites today use RSS as a "what's new" mechanism to attract users (King, 2001).

For the portal administrator, it's as easy as locating an RSS channel to point to via a standard HTTP URL on a given site and configuring the portal to intermittently grab a new copy of the file. Many popular general news sites such as Salon and CNN make their headlines available via RSS and advertise the fact—after all, it's another way to generate traffic to a site and reach an audience they might not otherwise hit.

Most sites will use an aggregation service to advertise their RSS channels. Several sites offer a variety of services ranging in price from free to thousands of dollars in licensing fees. Below are several examples:

> http://www.xmltree.com
> Claims to be the Web's most comprehensive directory of free syndicated content, though navigating the directory is somewhat difficult.
>
> http://www.moreover.com
> Provides syndicated content for a fee.
>
> http://www.newsisfree.com
> Contains thousands of channels, easy-to-use directory.

Choosing the Right External Channels

Because it's so easy to add external channels, it may be tempting to add every RSS channel that looks like it may be of interest to anyone in your organization. The portal's list could quickly balloon to dozens of sites that offer much of the same news—Fox Headline News will vary little from CNN Headline News, if only because both rely heavily on the Associated Press to provide the headlines. Criteria for selecting external channels include the following:

Appropriateness. Is the news source appropriate for its potential end users? Does the administration understand and condone employees using organizational resources to access this information? Headline news pertaining to genetic research may be acceptable, while football scores may not be. Institutional policies on acceptable use of Internet resources are an essential guide.

Reliability. Is the source reliable? Before selecting a channel to include, its origins should be clear. Is it a well-respected source, either nationally or a known resource to your users? As a provider of information, you have new responsibilities in ensuring the veracity of the content you've selected.

Currency. Is the source updated frequently? websites can fall into disrepair when their authors don't devote constant attention to them, and channels are no different. Watch for updates to a particular channel and be comfortable with its frequency. Make sure your content providers are dedicated to the channel services they provide, and not just experimenting or playing. Also important is synchronizing the time your portal looks for new copies of the channel from its source—many portal products will internally cache the content to improve performance, but you may want to override that or increase the frequency of cache updates if the channel content is time sensitive.

Options. Users of the portal can be overwhelmed if they have many sources to choose from. Do your users a favor by pre-selecting strong content providers. Provide a balance of options without making it laborious to scroll through menus of similar choices—two general news sources should be enough, for example, to provide a diversity of viewpoints.

Scope. Define the scope of what you want to provide your users before selecting channels. It's easy to get carried away with trying to provide a feed of channels from every Web resource the users might access. A channel that pulls sports scores for the day might be appropriate; articles detailing the latest baseball trades might only be distracting.

Specialize. Having several general news sites to choose from can be handy for the portal's users, but even more useful is a selection of sites that cater to special populations. Every department is full of specialists in a particular field—Human Resources, Networking or Purchasing—and there is likely a news channel targeted toward each field available among the thousands of RSS channels. If the portal can provide sources of news that were otherwise unattainable or unknown, the project will gain instant credibility.

Focus groups are key to providing specialized content. One important question to ask is what sites the members visit most often, without implying a judgment about their work-relatedness. Identify popular sites and customize your external channel list based on what people use most.

Leveraging Pre-Built Channels

Another way to add valuable content to your portal is to look to what others have developed. The more you can rely on others' good work, the more focused your resources can be on other important areas.

Many portal vendors provide a set of pre-built channels as part of a default installation of the product. Having a set of channels to work with immediately is invaluable. It gives you the opportunity to demonstrate the site without investing a lot of time and gives your developers some examples of how to begin creating their own channels. There should be no need to build from scratch such basic portal services as personal bookmarks, general announcements, hooks into e-mail systems or Internet searches. Some companies have even partnered with external news sites to provide licensed RSS news channels.

Indeed, one factor in deciding on a portal vendor should be what services it provides out-of-the-box. If you receive only a framework on which to hang applications, you'll spend considerable development time creating basic services and not on customizing the portal for your unique needs.

Another important factor in deciding on a portal vendor is what its plans are for developing new channels for its customers, and how it might facilitate communities of developers sharing custom channels they have developed. Some will be quick to provide you with custom channels created as add-ons by their consulting division for hefty fees, while others may facilitate forums of customers wishing to trade channels they have created. Still others might provide Application Programming Interfaces (APIs) to various services, which make known the necessary methods for connecting to other products and serve as building blocks for the production of channels, but do not necessarily provide the channel code itself. Questions to ask the vendor include:

- What is your schedule for delivering new channels?
- Do you take customer feedback in creating and prioritizing your channel development?
- Do your customers have to pay per channel as they are developed, or are new channels included in a maintenance contract?
- What channels have your customers created? Do you facilitate access to them?

The open-source uPortal product is philosophically built around the concept of sharing channels and source code among its users. This is one of its main appeals. Users can find channels in the Java in Administration Special Interest Group Clearinghouse website, at https://www.mis4.udel.edu/JasigCH/. Descriptions of the channels and contact information for the developers are available, but no actual channel code is provided. In this way each institution can preserve its individual licensing arrangements, charge fees for its work and/or offer varying degrees of technical support.

In addition to the portal vendor and communities of users, a third source of pre-built channels comes from individual vendors of the services you wish to enable through the portal. More than ever, those who sell solutions to higher education are being forced to enable their products through a variety of interfaces and share well-documented APIs to their products. This allows easy coupling of their services to a portal. Good examples are two leading course management systems, WebCT and Blackboard. At the time of this writing, both offer APIs to allow integration with third-party portal products, though additional fees may apply.

Even with APIs and out-of-the-box, functioning channels, don't expect to be able to make a useful portal based solely on services others have created. Such services can go a long way in providing basic functionality, but you won't realize the power of a portal until you make it work with your own business processes.

Creating Custom Channels

Incorporating external channels into a portal can require as little work as adding a URL to a database field. Pre-built channels from vendors require reading documentation and connecting the dots. For custom-developed channels, however, there is no surefire recipe. An institutional portal that does not incorporate custom channels is really no better than a commercial portal such as MSN, Yahoo! or Excite.

To truly benefit from a portal, the institution should use it as a one-stop fulfillment center for its most commonly used business processes. The portal must bring together in a common interface various back-end systems like human resources, student administration, financial systems, research and data warehouses, as well as homegrown applications built for the specific needs of the institution. It should come as no surprise that creating custom channels is the most complicated component of a portal and the most difficult of the three types of channels to work with.

Porting Existing Web Services

Those services that already have a Web presence must be revised to fit into the portal framework. Implementers need to consider several issues:

- *Authentication modules for existing Web applications will likely be removed and replaced with portal authentication services.* One of the advantages of a portal is its provision of a type of single-sign-on capability. Web applications that demand additional credentials due to security concerns (and not technical limitations of the specific application), should request authentication at the portal level, not at the individual channel level. This may require running the entire portal from a Secure Socket Layer (SSL) encrypted server.
- *HTML output must be redesigned.* Trim excess white space, graphics, and HTML presentation code. Large header graphics can be eliminated in lieu of simple text headers whose size and color are controlled by the portal's stylesheet. If possible, it's best to have a majority of application interfaces reside completely in a channel space within a portal page. This greatly changes the dimensions allowed, from 550-800 pixels wide to as little as 150-300 pixels.
- *Simplify not only the presentation but the functionality as well.* Any application that requires more than a few form fields such as radio buttons, pull-downs, or text boxes should be passed on to a separate screen. Consider offering a simplified version of the most common elements of your application in a channel and, from there, link to an advanced screen for more detail. The same goes for the output of an application, too. If more than a few lines, consider opening new windows rather than redrawing the portal screen. The ideal portal application is one that can extract a few lines of pertinent information from a source and present it to a user based on his or her login in a "dashboard" style, needing very little space on the screen and very little input from the user.

Transitioning Business Processes

More challenging than redeploying Web applications is transferring business processes from a non-Web environment into a portal framework. By far, channels like this will take the bulk of an institution's time and energy in the overall portal project, because the processes they embody are so likely to cross organizational and technical boundaries of all kinds and require in-depth analysis. Take a "My Benefits" channel as an example. You may want to include such features as:
- Earnings summary with federal, state, and local taxes detailed
- Annual salary amount
- Retirement plan details and savings amount
- Insurance information including the companies and types of insurance such as medical, dental, vision, and life

- Vacation and sick days taken
- Benefits announcements and reminders

While all six items above relate to a single individual's benefits, information about them can be stored in literally six or even a dozen different systems on campus and off. Calculations to produce earnings deductions may come from a centralized state authority; salary from a financial system; retirement funds and insurance plans each from one of several contracting companies; sick and vacation days from the databases of dozens of departments; and announcements from the Human Resources director's office.

Each silo of information has hard-working staff dedicated to the integrity and security of its system. They each have projects, deadlines and priorities, none of which are likely to be under your control. Be prepared to answer such questions as:

- How will information be extracted from an existing system? You will likely not get permission to have users directly access financial systems in real-time, for instance. Arrangements must be made for scheduled exports, creation of data views or establishment of data marts.
- What will users be able to do with the information? Will they be allowed to update their own information? If so, what specific fields?
- What security methods are you using? Most contributing systems already have authentication schemes in place—how will you get the existing method to work from within your portal? How will you protect sensitive information from others?
- How much of the contributing office's time will this take away from other projects? What's the organization's priority for it—does it take precedence over the department's current projects, and if so, by whose authority?

In developing interfaces to massive, enterprise-class systems such as human resources or finance, many additional factors come into play in prioritizing development and figuring out a timeline. When you plan on going into production with your portal, what version of your Enterprise Resource Planning (ERP) software will be in place? There are likely scores of programmers working at all times on patches and upgrades to such systems, and it is critical to understand where they will be in the system maintenance cycle at the time you are ready to interact with them. Providing Web access to their systems may not be a priority for them when they have a tight schedule to deliver salary changes by the next pay period, for example, which is why plenty of lead-time is necessary to get into their queue of projects.

Technical hurdles are seldom the most difficult aspect of interfacing to ERP systems. More frequently, difficulties and delays arise from attempts to cross cultural and political barriers. While critical to an organization, ERP systems are nearly never managed as open resources. The primary customers of a campus's multi-million dollar ERP system can literally be a few top-level executives who rely on reports from the system to make key business decisions. In such cases, allowing individual staff glimpses into such systems—to generate queries, run their own reports and even input their own data—requires a radical paradigm shift.

Until recently, Web developers generally created small, single-use systems with project turnaround times counted in weeks. What Web developer can natively 'speak' financial systems with hundreds of database tables, terabytes of information and version update cycles that stretch into years? Similarly, ERP programmers have often been isolated from the larger campus community and may be unfamiliar with rapid development cycles and Web interface issues. Working on this type of custom channel will require careful attention by a project manager and a cross-functional team of both portal developers and ERP programmers that understands and agrees upon the goals of the portal.

PRIORITIZING YOUR CHANNEL DEVELOPMENT

Once the channels you intend to incorporate into your portal have been selected, through committees, focus groups, process analysis or the results of a pilot, you must then prioritize that list based on your short- and long-term goals.

This is more difficult than reordering a simple list. Before you can begin grouping your channels into production phases, you need to determine the complexity of each channel's implementation. Several key points about each channel must be completely understood:

- What type of development work will be needed to complete the channel? Is it primarily reformatting an XML-based document or will it require encapsulating business logic into Java beans?
- What resources are available? Which programmers among the available staff can work on channel development? What percentage of their time can they devote to this, and do they have the necessary training to complete the tasks?
- To what extent does your project rely on information providers outside of your control? What priority does your channel development project have with them?
- Do interdependencies among the channels require a particular development sequence?

Once these questions have been answered, the development team can begin to create a timeline for the development of the channel and assign appropriate resources to it.

There are basically three levels of priorities in channel development: immediate, which means the channel is necessary to demonstrate to stakeholders the usefulness of the project; high priority, which applies to channels that are required before a portal can be launched in a production environment; and low priority, which applies to channels that can be built to round out the capabilities of the portal as the project matures. Being flexible in moving channel priorities will be a necessity as hurdles and opportunities arise.

Each of these priority classes is discussed below.

Immediate Needs

This is the set of channels that are critical to gaining acceptance with the portal's potential users and among the major stakeholders in the project such as executive-level sponsors or committees. These are the 'ah-ha!' pieces that help people understand what the project is all about.

One method of identifying these channels is simply to ask the question of a stakeholder, "What 'killer application' do you envision will make this project a success?" This can be a loaded question, and the likely response will be a difficult proposition or else it would have been solved before. A sample response might be, "a self-service Web interface to our mainframe-based student registration system." Even if the suggestion is unreasonable, at least it helps identify the direction, motivation and vision of the stakeholder. If a true self-service application is impossible in the first portal iteration, then perhaps displaying personalized information like a student's class schedule would be enough to satisfy initial goals.

To fully demonstrate the capabilities of a portal, examples of each type of internal channel should be included in the pilot rollout. This suggests incorporating several external channels such as weather and general news, porting one or two existing Web applications and developing at least one custom-created channel.

High Priority

Channels that provide basic portal functionality should be a high priority. Authentication services are a good example—while not necessary to demonstrate the portal to stakeholders, they are a key requirement for all other personalization and customization channels to be added in the future. Other important channels to include before launching might include announcements relevant to the user's role in the institution, personal bookmarks and role-based calendars.

Low Priority

Low priority channels may be important to the long-term success of the portal, but can be put off until after you launch your initial portal. These are channels that you'd like to have and would be useful in rounding out the portal, but don't affect basic functionality or relate to the initial goals of the portal. Some channels will have to be pushed back past the launch date for reasons beyond your control. An external system's upcoming milestone, such as being able to export in XML format, is one example.

Low-priority channels can add usability enhancements to other channels as well. Providing an RSS file editor channel for those responsible for updating announcements, for instance, could simplify the updating process for the publishers. Portal administrators would benefit from channels that provide traffic monitoring tools or user profile management.

Other channels could include portal administrative features like user profile management and traffic monitoring tools.

FUTURE TRENDS

Portal technology is still nascent, with traditional higher education enterprise vendors scrambling to fill this gap in their product offerings. Many 'pure-play' portal vendors—those companies whose sole product line revolves around portal services—and open-source consortiums of portal developers, both in and out of higher education, are still working on beta versions and prototypes.

Institutions with years of experience in portal development are extremely rare. Most have recently launched a portal, are planning to launch one within a year or are still investigating strategies. As more institutions see the advantages portals offer, they will seek methodologies for selecting and deploying portals, and creating content for them. Traditionally, institutions of higher education have worked collaboratively to define standards and share experience, code and strategies.

Some providers of portal solutions encourage and depend on this collaboration, while others have done little to foster such user communities in hope of generating additional consulting revenues. Given the low budgets and highly collaborative nature of higher education, portal development schemes that do not facilitate open APIs and facilitate shared channel development will ultimately falter in this market.

BRINGING IT ALL TOGETHER

This chapter has offered a way of organizing a channel development strategy by identifying potential content, classifying it and then prioritizing it into distinct categories. Effective ways of identifying content on campus include forming committees, conducting focus groups and launching pilot portals, all of which should be done to varying degrees according to what formula works best on a given campus.

In identifying content it's important to include representatives of the campus communities that will be using the portal, and ensure that they detail what they actually need rather than what they think they need. External channels must be appropriate to the institution and reliable. Understanding your portal vendor's philosophy in providing channels to its customers is vital. Customized channels that streamline your institution's business processes will be the most valuable parts of your portal, and the bulk of your portal development work will be in creating them.

Prioritizing your channel development requires understanding the complexity of each channel's implementation and what resources are available for the implementation, on the external teams that manage the systems with which the portal must interface, and on the portal development team itself. Grouping channels into immediate, high- and low-priority levels is key to strategically attacking your list of channels to be developed.

REFERENCES

Gleason, B. W. (2001). uPortal: A common portal reference framework. *Syllabus Magazine,* October. Retrieved October 10, 2001, from http://www.syllabus.com/syllabusmagazine/article.asp?id=4136.

King, A. (2001). The evolution of RSS. *WebReference.com*. May. Retrieved October 10, 2001, from http://www.Webreference.com/authoring/languages/xml/rss/1/.

Neilson, J. (1997). The use and misuse of focus groups. *Useit.com: Jakob Neilson's Website*. Retrieved November 14, 2001, from http://www.useit.com/papers/focusgroups.html.

Chapter VI

Campus Portal Strategies

David L. Eisler
Weber State University, USA

ABSTRACT

This chapter is designed to assist campuses and their leaders in determining whether to pursue a portal project. For those universities that choose to create portals, a series of strategies and approaches are presented to guide and assist in the success of the effort. This material is provided from the perspective that campus portals can provide a new way to connect with students, faculty, alumni and the community. Strategies are presented to determine whether to undertake a portal project, and the campus readiness for this effort. Nine different approaches to campus portals are presented, together with suggestions on project organization. Project success factors are developed together with potential planning pitfalls for campus portal projects. Finally future approaches for portals are discussed with thoughts for portal acceptance on campus.

Copyright © 2003, Idea Group Inc. Copying or distributing in print or electronic forms without written permission of Idea Group Inc. is prohibited.

When they burst on the campus scene, portals were seen as the next "killer application" for information technology in higher education. With portals, universities would create seamless interfaces coordinating electronic information for everyone on the campus and beyond: students, faculty, staff, alumni and the community (Gnagni, 2001). Perhaps most amazingly, this new wonder technology could be provided free to campuses by vendors willing to construct these interfaces in exchange for "click through" revenues.

With the perspective time provides, it is now easier to separate solid, reliable technology efforts from some of the initial high flyers in the campus portal market that provided more style than substance. Just as "dot.com" was replaced in the national tech economy by "dot.bomb," some high profile portal vendors no longer exist. Today there are wonderful working examples of campus portals and encouraging initial reports on the adoption and usage of these interfaces by members of the university community. Although less publicized, there are also examples of well-intentioned portal projects that have collapsed and failed. For campuses contemplating a portal project, the question looms whether this is another example of a failed technology that promised more than it could deliver, or if portals are an important innovation campuses should pursue as part of a balanced technology strategy.

This chapter is designed to assist campuses and their leaders in determining whether to pursue a portal project. For those universities that choose to create portals, a series of strategies and approaches are presented to guide and assist in the success of the effort. This material is provided from the perspective that campus portals can provide a new way to connect with students, faculty, alumni and the community. While still in an evolutionary state, portals are an interface between colleges and constituent groups that can become new mechanisms to organize campuses, and offer the promise of new ways to create communities of learners. Against this backdrop and in a time of economic uncertainty, it is extremely important for colleges and universities to consider portal adoption and implementation as carefully and seriously as any other large-scale strategic effort.

THE CAMPUS PORTAL

A campus portal may be defined as a single integrated point for useful and comprehensive access to information, people and processes. While portals have a rapidly evolving set of features and characteristics, they can be described as both personalized and customized user interfaces providing users with access to both internal and external information. Campus portals provide the opportunity to create:

- Gateways to information
- Points of access for constituent groups

- Mechanisms for communication
- Community and learning hubs (Eisler, 2000)

Perhaps the best way to understand some of the possibilities of campus portals is to visit and experience them. A good place to begin is with portals like UCLA's "My UCLA," the University of California, Davis's "My UCDavis," the University of California, Irvine's "SNAP" (Simple Navigational Administrative Portal) and the University of Washington's "myUW." All provide a high degree of integration with university systems and permit guest access.

Portal Organization

As illustrated in Figure 1, portals conceptually consist of three basic components. The center circle represents the functions that provide the user access, identification and security. The outside ring of circles represents content modules, user tools or data resources. The number and richness of these features can be increased or decreased depending upon the intended user group, the degree of functionality desired and the integration of administrative applications. Finally there are connective pathways that convey requests to the system and information or processes to the user (Eisler, 2001).

Figure 1.

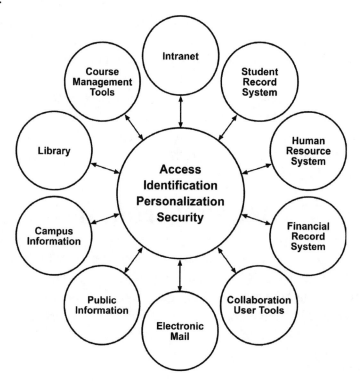

An effective campus portal requires the coordination of a variety of components and systems to provide the access, information and interactivity desired. Campus portals provide access to university resources and provide security for university data. A wide variety of information can be accessed by or pushed to campus portal users. This information can be selected by users, be part of a person's records or be directed by the university to the individual.

Campus portals can provide a wide variety of work tools or access to them, in this process becoming a personalized desktop for users. Communications and interactive functions can be a part of the portal infrastructure or accessed through it. Effective portals provide access to and can enable university e-business functions. Using a university intranet, a portal can provide access to data, applications and forms through the employee's desktop.

Faculty portals can provide management tools for both Web-enabled and on-campus classes. Portal-enabled systems can become important mechanisms for communicating with students, continuing classroom discussions and encouraging interaction outside of class.

Considering this complex list, it is easy to imagine the value and functionality a well-designed portal effort can provide, and to understand the potential technological challenges portals can represent for campuses. (For additional information on portal functionality, see Box 1.)

Box 1.

Portal Functionality and Features

An effective campus portal requires the coordination of a variety of components and systems to provide the access, information and interactivity desired. Campus portals provide access to university resources and provide security for university data.

- **Gateway**—The system identifies approved users through a single sign-on procedure.
- **Security**—Users are allowed access to information they can see, to change information they can change and no more. Those who should not see or change information are denied access to it. A wide variety of information can be accessed by or pushed to campus portal users. This information can be selected by users, be part of a person's records or be directed by the university to the individual.
- **Customized information**—Users receive information for or about themselves and their activities. In the case of a student, this might be a class schedule, a degree checklist, bill balances or a reminder that a library book is due.
- **Channeled information**—Portals provide information from internal and external sources. Information examples include weather, news, entertainment, stock information, campus sports and newspapers.
- **Pushed information**—Universities, and users to a lesser extent, can select the information individuals and constituencies should receive. Examples include announcements or e-mails to all or groups of students, to particular majors or to students with specified interest; university deadline reminders; campus calendars; social announcements; and student activity notices.

Copyright © 2003, Idea Group Inc. Copying or distributing in print or electronic forms without written permission of Idea Group Inc. is prohibited.

Box 1. (continued)

> Campus portals can provide a wide variety of work tools or access to them, in this process becoming a personalized desktop for each user.
> - **Personalization**—Users can edit their portal page's look, features and arrangement, and at least some sources of information available on it.
> - **Internet tools**—Portals can provide search and navigation engines for the university intranet, university Web pages and the entire Internet. They can include tools to save favorite websites, create homepages, and create or post to message boards.
> - **Personalized tools**—Users can maintain integrated planners and calendars, create to-do lists and Web homepages.
> - **Library tools**—Users can create their own access to the campus library, on-line bibliographic resources, and databases through a MyLibrary function.
>
> Communications and interactive functions can be a part of the portal infrastructure or can be accessed through it.
> - **Interaction**—Portals can provide interfaces to chat, e-mail, address books, threaded discussion lists, listservs, message boards and bulletin board postings.
> - **Schedule management**—Portal interactivity can extend to coordinated calendar functions with the ability to search multiple schedules and create meeting times for university members, to export this information to multiple devices and to remind users when it is time for meetings.
> - **Electronic balloting**—Campuses may use portal technology for voting, survey functions and to poll campus constituents.
>
> Effective portals provide access to and can enable university e-business functions. Using a university intranet, access can be provided to data, applications and forms through the employee's desktop.
> - **e-Business**—Portals provide integration into university back-office operations, a one-stop interface for educational transactions over the Web and connections to the bookstore for shopping.
> - **Workflow and application integration**—Staff and faculty can access data and applications needed to do their work in a real-time environment, create personalized data reports and use contact or project management systems.
>
> Faculty portals can provide management tools for both Web-enabled and on-campus classes. Portal-enabled systems can become important mechanisms for communicating with students, continuing classroom discussions and encouraging interaction outside of class.
> - **Course management**—Portals can provide course management tools or integrate with existing systems.
> - **Course communication tools**—Faculty can send class announcements, access discussion lists and listservs, collect and grade homework and assignments, and post grades electronically.
> - **Discipline-specific resources**—Using portal capabilities, course sites can be extended into vertical portals for discipline-specific information.
>
> Considering this complex list, it is easy to imagine the value and functionality a well-designed portal effort can provide, and to understand the potential technological challenges portals can represent for campuses.

SHOULD OUR CAMPUS PURSUE A PORTAL PROJECT?

It is likely that most institutions will at some time consider a portal project. Portal discussions and implementation projects should collaboratively involve the campus community in developing answers to three simple questions: Who?, What? and Why?

- Who are the intended users of your portal?
- What will the portal provide your users?
- Why are you doing a portal?

Experience a Portal

Today campuses choose software for word processing, spreadsheets and e-mail based on use and experience. Colleges and universities create new designs and structure for websites by mining the vast amount of background on the Internet. The selection of these programs and the design of these efforts are based upon and shaped by personal and campus experience using these technologies. Imagine if a group on campus were asked to select one of these software products without having used it, having only read about it or experienced it only through a vendor demo. Unfortunately for many campuses this can be the atmosphere in which a portal project is designed, by user and technical groups whose experience with portal technologies is limited or nonexistent. It often happens that people who have not used portals regularly will develop proposals for portal projects and will make decisions about them. The campus knowledge of portals will most likely be limited to online visits to one or two portals, the study of written materials and observation of vendor demonstrations.

Before deciding on a campus portal project, it is beneficial if those involved can actually configure and use a portal. John Ellis (2001) captures this approach in the words of Yale economist, Robert Shiller: "If you really want to understand something—really understand it—don't just analyze it, do it." This is possible for even the most non-technical user, and the insight developed will be invaluable. Begin by configuring a personal portal using one of the free commercial services. Excite, Netscape and Yahoo are examples of companies that popularized configurable personal portals (Jacobsen, 2000). An especially interesting example is provided by Yodlee, which allows the user to build a financial dashboard accessing finances, e-mail accounts and a large variety of news sources.

After choosing a portal, take the time to understand the options available, configure a portal to your personal interests and tastes, and commit to using this portal as your homepage for at least a month. This experience will provide useful insight into the value a portal can provide aggregating content on the desktop. In time, the portal will become more than a homepage: it can become a personal information and access center.

For frequent Internet users the transition to a portal will require changing what may have become intuitive habits in terms of Internet use. While the differences in interface and page layout will create an adjustment and learning curve, if the user visits the same Web pages regularly for updated information, the advantages of a portal will soon become evident. Intrigued with this feature, some users will add more and more functionality until the page becomes large, slow and cumbersome. For these power users, multiple or layered portal pages may be the answer. As users gain experience with the portal, they should assess the value of the portal experience. What does the personal portal provide that they did not receive from their homepage? If the portal page provides increased convenience, functionality and access, it is extremely unlikely the user will want to return to his or her prior homepage.

Using a portal will also provide experience with what can be the frustrating limitations of a portal. Limitations in layout or display will quickly become annoying. For example, if only a portion of a page of personal bookmarks is displayed, going to a second page for access to regularly visited sites will quickly become an annoyance. While commercial portals provide the ability to choose and organize content on the page, users are frequently not able to size or resize content windows. Portal applications require bandwidth; if the site provides slow response, this will discourage use, especially over traditional modems. Finally, managing a portal requires patience and thoughtful effort if the tool is to evolve in effectiveness.

In addition to a personal commercial portal, project members should consider using an existing campus portal that permits guest access and configuration. Two examples that allow this are the University of Texas's "UT Direct" and Yale University's "Yale Station." Once this portal has been selected and configured, the user should again commit to using it for a time as his or her homepage.

Some important lessons can be learned from this personal experience with portals. For people who regularly access information from multiple sources, a personal portal will save a significant amount of time. The initial configuration can be time consuming, but as it is perfected, over time a portal becomes a valuable personal productivity tool. A personal portal is much more than a Web page: a well-constructed portal can be a "killer app" in the higher education setting, taking information access to the next level by organizing and integrating in one place information from sources that previously had to be accessed individually.

Additionally, it is instructive to construct a list of available Web-enabled sources of university information. It will quickly be evident if enough personalized services for students, staff or faculty are available via the Web to make a portal project viable. If the amount of personalized data users can access from university systems turns out to be limited, and access to data sources will not be improved by the portal project, campus energies might better be concentrated on creating access to these data and functions first, then considering development of a portal at some future time.

Certainly an institution can force users to adopt a portal, by making registration functions, grades, payroll stubs and other essential processes available only through the portal. The approach suggested here is a more interesting challenge: to see if users will adopt portals because of the increased functionality and productivity portals provide. It also reflects the spirit of higher education often embodied on campuses. This approach is best undertaken with a portal product that provides significant levels of customization and personalization.

CREATING A CAMPUS PORTAL

Two approaches to portal projects appear to have been successful on university campuses. The first identifies a portal project as an announced strategic technological goal. In this instance, successful portal projects can best be achieved through the combined efforts of faculty, students and academic support professionals. Deciding whether or not to have a campus portal and successfully implementing one requires effective cross-campus participation and commitment. The process begins by assembling the right group of people and having them address the right questions. The interactions of this collaborative group can create precisely the right environment to consider some of the questions critical to the successful development and deployment of campus portals. It will help campus representatives decide who will control what users see and can access. The type of cross-divisional institutional team required to create a successful portal might include the following people:

- Chief academic officer
- Dean or department chair
- Faculty (representing those who have adopted technology and those who are less experienced with it)
- Academic support professionals
- Chief information officer
- Information technology (IT) support professionals
- Library
- Student services
- Bookstore
- Students
- Prospective students
- Alumni

The second approach to campus portal projects begins as a more modest effort. In this a particular portal technology is chosen and small "expeditionary" groups are created to test the technology and create the initial portal implementations. This approach can avoid the creation of unrealistic expectations, allows initial failures to affect small numbers of users and can allow actual campus experience with the portal to shape the product rolled out to larger groups of users.

No matter which approach or combination of approaches is chosen, the process can assist each area of the campus to answer the following questions for the different categories of portal users:
- What options are available?
- What information is included?
- What services are offered?

In addition, the implementation group should consider answering such critical questions as:
- How can a campus portal extend, expand and increase the participation and communication among members of the campus community?
- How will the campus community accept, adopt and use the portal?
- What are the potential risks and problems associated with the portal?
- What is the value of a portal, for both the individual and the university? (How will that value be assessed?)

Box 2.

Portal Project Design Considerations

Features
- What functionality do we want for this project?
- What content will users receive that they cannot already access?
- If the content is already accessible, will the portal make access more convenient?
- How much choice will users have in the content they receive?
- Will the content keep users on the portal or send them to information sources off the website?
- How intuitive is the interface for users?
- Are we willing to have advertisements? If so,
  o How many?
  o Placed where?
  o Will we restrict ad content and advertisers?

Integration
- Course management system
- Existing databases
- Transactions
- Wireless
- Smart card

Technology
- Will our infrastructure support this?
- Can we provide single sign-on capability for users?
- Can we develop and/or connect to the following?
  o Access to student information
  o E-commerce functions
  o Speed of response and access
- Can we resolve security and data protection concerns?

Technical Considerations
- Hosting
- API (Application Program Interface)—ability to pass information to other applications

---

Box 2. (continued)

- LDAP (Lightweight Director Access Protocol)—allows user to query database via Internet
- User definition capability
- Custom information channel definition
- ADA (Americans with Disabilities Act)—accommodations for users with special needs
- Hardware requirements
- Pricing
- Vendor viability

Support
- Can we support it? Keep in mind that a campus portal is not a short-term venture. If it succeeds and is accepted by the campus and alumni, the portal will be something needing support, maintenance and upgrades long into the future.
- How many staff are needed, for rollout, for maintenance?

Content Support
- As a dynamic medium it will include a variety of content providers and will require more support than a static website. Support includes providing reliability 24 hours/day, 7 days/week.
- Who creates and updates content?
- Who controls content?

User Support
- Who provides, maintains online help, documentation and training?
- When is live help available? What is available 24/7?
- Who manages and maintains users?
- How easy is this to use?
- Is the product appealing to users?

Project Rationale
- Has there been a clear discussion and agreement on the benefits of a portal?
- Who understands and supports the project?

Budget
- How much funding is available? Ongoing? One-time?
- Do the potential advantages of a portal justify the commitment of people and resources for the project?

Analysis
- Is there a reasonable chance for project success, maintenance and growth?
- Will the campus community use a portal?

The implementation group should take time to think carefully about the design and projected capabilities for the portal. (See Box 2 for a list of portal design considerations.) Numerous examples of case studies from campuses that have created successful portal projects are available. These should be studied carefully in the development of a project and plan.

Portal Readiness

After considering the above questions and if there is still strong interest, support and motivation to create a portal, the implementation team would do well to step back for a moment before beginning and consider if this is the appropriate time to begin a portal project and whether the campus is ready to do so. Here are some signs that a campus may not be ready to begin a portal project.

- Network problems would prevent effective and equitable implementation campus-wide.
- Significant retention problems with technical support personnel would compromise the robustness of the portal services.
- Data sources, calendaring or student support systems require hand maintenance by a variety of offices.
- The campus student information system is not supported by a portal vendor or is not accessible by the Web.
- The campus lacks a spirit of collaboration on IT issues.
- The campus lacks consensus about the appropriateness of certain content (political, religious, advertising, etc.).

Another effective test of a potential portal is to demonstrate the technology to members of the campus community unfamiliar with the technology. If these demonstrations are met with a significant level of uncertainty as to what the portal is or the value it may bring, it may suggest that your core project group is not prepared to communicate the project to the campus. Project members can become evangelists for a portal, seeking to influence or change the opinion of those less certain of the need. If the potential value of portals is not clear to focus groups of potential users, it may be that the implementation is on a path of providing an interesting technology, but with limited functional use and faint hopes of widespread adoption.

CHOOSE A PORTAL STRATEGY

Campuses may choose a number of strategies in creating portals. The following categories suggest possible adoption paths and provide examples of successful, accessible campus efforts.

Build Your Own

A university can choose to build its own portal. Many of the early prominent examples of campus portals, like "My UCLA," and "my UW" from the University of Washington, are examples of university-developed portals. The creation of a portal in-house allows a campus to design and truly customize a portal to meet campus needs, systems and culture. It does however require significant technical expertise and resources.

At many colleges and universities, the information technology staff members are already spread too thin or lack the technical expertise to take on additional programming tasks of this complexity and magnitude. As a result, building a portal from scratch will be well beyond the capability and capacity of most institutions. Homegrown portal efforts also may create greater continuing costs for development and support. As the campus portal market becomes more mature, it will become increasingly difficult for individual campuses to keep up with new innovations and features. This could lead to a situation requiring a transition from a homegrown to vendor-based solution. For users accustomed to a system created according to their needs, this can be problematic.

Working examples of campus-developed portals include "Blink" (UC San Diego), "MyUB" (University of Buffalo), "MyUCDavis," "MyUNIverse" (University of Northern Iowa), "UTDirect" (University of Texas) and "Yale Station."

Partner With Others

A campus may choose to partner with others to create a shared code approach to portals. The JA-SIG group is a partnership of university and college campuses involved in this effort. JA-SIG is an acronym for Java in Administration Special Interest Group and the combined product is referred to as uPortal (Gleason, 2001). The first working JA-SIG portal was the University of British Columbia's "My UBC."

This shared approach seeks to lower the development investments of campuses by dividing the development of project segments among them. This shared source code strategy is not unlike the work behind the open-source Linux operating system. However, it is disappointing that some of the original lead universities in this effort have not moved beyond demonstrations to produce working portals for their campus using the JA-SIG technology. Like a "Build Your Own" approach, universities should not choose to "Partner with Others" without strong support, leadership and commitment from the information technology division.

An excellent example of a JA-SIG portal is the University of California, Irvine's "SNAP" (Simple Navigational Administrative Portal). Other working examples include California Polytechnic State University's "myCalPoly," Denison University's "myDENISON" and Laurentian University's "LU's Student Portal." Others universities with demonstration sites for the uPortal include Althabasca University, Boston College, Columbia University, Princeton University, the University of Delaware, the University of Switzerland-Geneva, the University of Hawaii, the University of New Mexico, the University of Nevada and Yale University. Interactive Business Systems is a vendor that works with campuses on uPortal implementations. Campus Pipeline has also announced plans to partner with the JA-SIG group.

Work with a Vendor

Just as in administrative student information systems, universities can work with vendors to create campus portals. These products provide differing levels of personalization and customization, and may require that university data reside on the vendors' servers. Combinations of cost and student data access are strong reasons to consider using a portal vendor. This is especially true for campuses whose student and/or administrative software systems are supported by a portal vendor. These vendor-based integration efforts show significant promise for solving problems associated with data access. The vendor approach can provide a reasonable and cost-effective solution, however the decision becomes more difficult when there is no access to student data or where a customized interface to these data has to be built when the vendor approach will not yield access to student data.

Many current portal products limit users to predefined channels of information. While this provides ease of management, it also works against the freedom of choice and access many users expect from the Internet. An exciting development in portal technology is a feature that allows users to create their own channels of information by capturing the URLs of favorite Web pages, dissecting those pages into content fragments and then reassembling the fragments on their own portal pages. Before ceasing to exist in the fall of 2001, Octopus.com delivered this level of personal choice. To a lesser extent, this same concept is embedded in Microsoft's Digital Dashboard offering (Hodder, 2001).

Dealing with portal vendors requires the same caution associated with the selection of major campus software systems. The portal market has developed rapidly. Many of the products and their features, like the companies that provide them, are still evolving. It is important to:

- Differentiate between functionality that is developed and that which is promised in the future.
- Be wary of vendors who may promise more than they can deliver. Like administrative systems, portal applications are offered to campuses by sales people whose interests may be motivated more by quick profits than long-term, mutually beneficial relationships.
- Pay less attention to product demonstrations and more time investigating the vendor's track record for providing support.
- Check with other institutions that have implemented the software, both those that have been recommended by the vendor and those that have not. Their experiences can be excellent indicators of the strengths and weaknesses of the vendor.
- Investigate the funding and stability of any portal company you might choose. In the vendor market some firms have disappeared, some have reworked

business plans, others have been bought up by larger companies desiring their technology or partnered with system software vendors. A wonderful product will be of limited or no use to your campus if the vendor is no longer in business or able to provide support.
- Be extremely careful of campus portal vendors who propose funding mechanisms based on click-through revenues. Just as in other dot.com sectors, this has not proven to be a sustainable business model. The cases of Mascot and zUniversity, both defunct campus portal vendors, are worthy of careful consideration as examples of how this strategy may not work.

A growing number of campus portal vendors are either administrative software companies or are partnered with one. These include Campus Cruiser (DataTel), Campus Pipeline (SCT), Jenzabar (CARS, CMDS, Quodata, Campus America POISE), ORACLE, PeopleSoft and SAP. Another important emerging trend is the development of portals that provide or support courseware management systems or a partnering arrangement. Examples of this include Blackboard, Campus Pipeline (WebCT) and eCollege. Other campus portal vendors include CNAV, FirstPerson.com and StudentOnline.

Use a Business Portal Solution

Some campuses may choose to use portal software developed for the private business sector. This can provide a robust product, especially for campuses that have homegrown administrative systems. Business portal solutions may prove expensive and require special adaptation of both the system and the campus data. It is an approach best undertaken with the guidance of an independent consultant well versed in the area.

A significant number of vendors offer specialized business portal solutions. Some leading companies in this area include Brio, Computer Associates, Corechange, Documentum, Epicentric, Plumtree and Viador. Worthy of consideration in this area are the efforts of large established firms including IBM's "Enterprise Information Server" and Lotus's "K-Station," Microsoft's "SharePoint Server" and "Digital Dashboard," Novell's "Portal Server" and Sun Microsystems's "iPlanet."

Create a System Portal

Some university systems are working with other higher education institutions to create portals. Most often these efforts focus on creating a common portal for prospective students, transfer students and their parents. As such, the portals are not designed to provide services for many in the current university community, but do provide a common data set and interface for potential college students. The combined resources of a state system can make this approach cost effective. The ability to apply to multiple campuses from a single website can prove attractive for prospective students.

The vendor product most often used for system portals is "Mentor" from xap.com. Mentor has been adopted by California Colleges, California Community Colleges, the California State University system and systems in Connecticut, Delaware, Georgia, Illinois, Kentucky, Massachusetts, New York, North Carolina, Pennsylvania, Tennessee, Texas, Wisconsin and West Virginia.

A different example of a state system portal implementation is "OneStart." Developed for the eight campuses of the Indiana University system, this portal links the campuses' general and campus-specific Web offerings and provides access to admissions for each campus, together with information resources.

Choose Specialized Portal Solutions

Campuses may choose to adopt specialized portals for specific services. One example of this is the myLibrary concept developed at North Carolina State University. With this the user configures a specific grouping of library functions and resources. While providing a great deal of user convenience, this approach is considered problematic by some academic librarians, who fear that students may not make full use of the wider range of library resources available (Buchanan, 2001). "MyLibrary" is a shared open source solution available to campuses without charge. In addition to North Carolina State, working examples include Virginia Commonwealth University, medical libraries at New York University and Southwestern Medical Center at Dallas, and "Your Library Web Page" from the University of Tennessee at Chattanooga.

Another specialized application is designed for alumni and athletic associations. These portals primarily deal with off-campus users.

Specialized portals can be selected from a variety of sources. Rather than adopting a single portal approach, some campuses may find that a variety of specialized portals meet their needs more efficiently. For example, a campus might choose a student portal developed by its administrative software company, a distance learning portal from its Web course management vendor, the "MyLibrary" approach to library resources and an alumni portal from a specialized vendor. While the integration of products from diverse vendors creates additional complexities for maintenance and support, these multi-faceted projects could prove effective.

One of the dangers campuses can face is the desire of constituent groups for their own independent portal solutions. This potentially places at risk the single-sign-on convenience portals are designed to provide. In addition it raises the question of compatibility of these potentially uncoordinated efforts, and the added maintenance and support they may require.

Develop an Interim Solution

For campuses in a period of technology transition, for example because of a planned or active migration to a new administrative software system, an interim or partial portal strategy may be the best approach. If the data sources on which the portal will rely have yet to be developed or may change significantly in the immediate future, it may be imprudent to invest in their integration into a portal.

During such a transitional period, it may make sense to adopt an interim portal strategy that does not provide everything a campus desires all at once. A gradual portal implementation can introduce users to portal technology at a limited cost while allowing limited technical resources to be directed to other more pressing prerequisite information technology needs. A drawback to this approach is that the portal may provide limited value to users, eroding campus support for future adoption of a full-featured solution.

Extend User-Specific Web Pages

A different interim strategy for campuses may be to extend the functionality of existing Web pages for campus groups. Solicited feedback from users can help to fine-tune these efforts to meet high-priority needs. While not providing the unified access, security and personalized features of a portal, this may prove a cost-effective alternative.

The difficulty with this option lies in the lack of a truly customized solution for users. An effort to address users' portal-related needs without deploying a true portal will most probably create a large page with much information or a Web approach that embeds information several pages down. Successful implementations of this approach include the California Institute of Technology's "CalTech Portals" and the University of Virginia's "ITC."

Choose to Not Do a Portal

Finally it may be that a college or university does not need a portal, or has more important challenges on which to focus funds and technological resources. This is especially true when Web access to campus data and services is limited. If a portal project could not provide this access, or if other limitations would make adoption unlikely, efforts would be better directed at creating an intranet instead.

PORTAL PROJECT SUCCESS FACTORS

A critical factor for portal success is access to data and systems. As a result, the institution's student administrative software system will be a major determinant in portal selection and development. Is the student data system a commercial product or a homegrown effort? For the former it may be most feasible to work with the system vendor's portal or portal partner. For home-developed systems it is unlikely current vendor solutions will interface with student data without significant adaptation. Portal access to student data is simplified if this information is already accessible via the Web.

A second critical portal factor can be the courseware management system used on campus. If the campus has made significant progress in the adoption of a courseware management system, this also should be carefully considered in the selection process. Again it may be most feasible to work with the system vendor, assuming it provides a portal product. Faculty adoption of a portal will increase if the portal functions as an interface to the courseware system.

Information technology staff resources are critical to the success of portal projects. Many IT departments are understaffed, undertrained and overworked. A portal project will be a significant addition to IT staff workload. Does the capacity exist not only to develop, but also to support and maintain the project? If not, will the effort be outsourced or will additional staff be hired to support the project? Given the complexity of most portal efforts, understaffing the project can be an invitation for disaster.

What is the budget for the project? Budgetary support is a significant indicator of campus and administrative support for portals. The absence of strong central support is a danger sign that the portal project may not be understood, or campus need for a portal has not been demonstrated or perceived. Budgetary resources will need to be both one-time, for software, servers, programming, implementation and promotion, and continuing, for user support, maintenance, expansion of effort and content creation.

With all the attention directed to campus portals, it is possible to get caught up in the hyperbole surrounding this topic. Campuses need to make certain that their constituencies both need and will benefit from the project. Ultimately, campuses must determine what they are able to do—not only whether they have the wherewithal to undertake a portal project, but also why they want to do so.

Planning Potholes

Portal planners should set modest goals for the first campus portal implementation. The more features or bells and whistles, the more likely it is that the project will become overly complex, the delivery date will be delayed and the resulting product will be flawed or unreliable. The planners should determine what key features are needed for the portal to succeed, begin with a modest list of goals and fight against project creep. In many cases simpler efforts will be wiser, more easily completed and more reliable. Additional features and capability can be added over time. It is a prudent approach to promise less and deliver more.

Portals do not improve on information; they only present it. A portal will only be as good as the information it contains. The design of a portal should focus as much on the information the portal is to contain and the development of this information as on the technology of the project.

The management of user expectations can be a difficult task. Different categories of users and different users within a category will desire different functions. Consider carrying out a pilot project for a limited set of users. This can prevent major embarrassment and rejection of larger scale campus portal efforts. It is extremely important that the portal project develop a track record of success and reliability.

The purpose of a portal project is to provide additional functionality for users. It would be fatal to the project to forget, ignore or bypass users in the design process. Projects can develop lives of their own and the original purposes can be lost. Portal projects must begin with significant user input. With proper selection and involvement of potential users, the design team can develop a project that will meet broad campus needs.

Many people on campus do not know what a campus portal is and as a result can see no reason for one. They may see the resources proposed for a portal project as better used on other campus priorities. It is important not to underestimate potential campus resistance to the project. Ultimately the portal will need to facilitate increased access and provide additional functionality if users are to adopt it. No amount of selling or influence will change behavior if functional value and ease of use are not present in the product.

THE PATH AHEAD

While it is easy to conceptualize the potential value of portals, to date limited published information is available about student acceptance and use, and even less about faculty and staff acceptance. Clearly, functionality and value will be key concepts in user acceptance of portals. In time, portals can become increasingly user-friendly as they track, understand and adapt to user actions and access patterns. In the future, higher-education portals may self-assess, reconfigure, anticipate needs and requests, and make recommendations to users, much as commercial sites now recommend choices based on visitors' usage patterns.

As portals are integrated into wireless technology, more will include the ability to use synthesized speech and deliver information in this format to users' cell phones. For the present, campus portals offer the promise of an individualized view of information and increased access to resources. In the fabric of today's college experience, where campus life is often fragmented, the potential ability of portals to extend the academic experience beyond the classroom to create 21st century learning communities is an exciting and fascinating possibility.

Ultimately, portals will not create community; people create community. The success and future of campus portal projects will be a direct reflection of the people involved and the collaborative efforts they are able to establish. Campus portals provide a wonderful opportunity to extend and enrich the education that colleges and universities provide. The *process* of creating a portal is among the first stages in creating the new communities that higher education can become. Collaborative processes can help to ensure that the planning and development of portal projects are both inclusive and reflective of your campus.

REFERENCES

Buchanan, E. (2001). Ready or not, they're here: Library portals. *Syllabus,* 14(12), 30-31.

Eisler, D. (2000). The portal's progress: A gateway for access, information, and learning communities. *Syllabus*, 14(2), 12-18.

Eisler, D. (2001). Campus portals: Supportive mechanisms for university communications, collaborations, and organizational change. *Journal of Computing in Higher Education*, 13(1), 7.

Ellis, J. (2001). What is the new economics? *Fast Company*, 50, 118-124.

Gleason, B. (2001). uPortal: A common portal reference framework. *Syllabus*, 14(12), 15.

Gnagni, S. (2001). Portal quest. *University Business*, 26-31.

Hodder, S. (2001). Dashing ahead: customizable Web parts make Microsoft's digital dashboard a unique portal. *University Business*, July/August 68.

Jacobsen, C. (2000). Institutional information portals. *EDUCAUSE Review,* 35(4), 58-59.

ADDITIONAL RESOURCES

A collection of materials about campus portals is available at http://weber.edu/deisler/portal.htm.

Anders, G. (2001). Inside job. *Fast Company,* 50, 177-184. Retrieved April 26, 2002, from: http://www.fastcompany.com/online/50/bestpractice.html.

Boettcher, J. & Strauss, H. (2000). What is a portal, anyway? *CREN TechTalk Series*. January. Retrieved April 26, 2002, from http://www.cren.net/know/techtalk/trans/portals_1.html.

Choden, A. (2000). A hitchhiker's guide to learning portals. *Suite101.com*. Retrieved April 26, 2002, from: http://www.suite101.com/article.cfm/training_and_development/41704.

Eisler, D. (2001). Selecting and implementing campus portals. *Syllabus*, 14(8), 22-25. Retrieved April 26, 2002, from: http://provost.weber.edu/Syllabus/Campus%20portals.wpd.

Geith, C. & Wagner, C. (2000). Preparing for campus portals. *CREN TechTalk Series*, March. Retrieved April 26, 2002, from: http://www.cren.net/know/techtalk/events/campusportals.html

Gilbert, S. (2000). Portals demand collaboration—can portals support it? TLT Group, August. Retrieved April 26, 2002, from: http://www.tltgroup.org/gilbert/SyllabusCol2.htm

Katz, R. N. (2000). It's a bird. It's a plane. It's a ...portal. *EDUCAUSE Quarterly*, 23(3), 10-11. Retrieved April 26, 2002, from: http://www.educause.edu/ir/library/pdf/eq/a003/eqm0038.pdf.

Kidwell, J., Linde, K. & Johnson, S. (2000). Applying corporate knowledge management practices in higher education. *Educause Quarterly*, 23(4), 28-33. Retrieved April 26, 2002, from: http://www.educause.edu/ir/library/pdf/EQM0044.pdf.

Looney, M. & Lyman, P. (2000). Portals in higher education: What are they, and what is their potential? *EDUCAUSE Review*, 35(4), 28-36. Retrieved April 26, 2002, from: http://www.educause.edu/pub/er/erm00/articles004/looney.pdf.

Loshin, P. (2001). Single sign-on. *Computerworld*, February 5. Retrieved April 26, 2002, from: http://www.computerworld.com/cwi/story/0,1199,NAV47_STO57285,00.html.

Norman, M. M. (2000). Portal technology: Into the looking glass. *Converge Magazine* (Suppl.). Retrieved April 26, 2002, from: http://www.convergemag.com/SpecialPubs/Portal/portal.shtm

Olsen, F. (2000). Institutions collaborate on development of free software. *Chronicle of Higher Education*. May 5. Retrieved April 26, 2002, from: http://www.chronicle.com/free/2000/05/2000050501t.htm.

Paadre, H. & King, S. (2001). *College of the Holy Cross: Electronic Community and Portals*. White paper retrieved April 26, 2002, from: http://www.mis2.udel.edu/ja-sig/holycross.doc.

Pittinsky, M. (1999). Campus and course portals in 2015. *Converge Magazine*, October. Retrieved April 26, 2002, from: http://www.convergemag.com/Publications/CNVGOct99/Possibilities/Possibilities.shtm

Sistek-Chandler, C. (2000). Portals: creating lifelong campus citizens. *Converge*

Magazine, October. (Suppl.). Retrieved April 26, 2002, from: http://www.convergemag.com/SpecialPubs/CampusCitizens/defining.shtm.

Steinbrenner, K. (2001). Unlocking ERPS with portals. *EDUCAUSE Quarterly,* 24(3), 55-57. Retrieved April 26, 2002, from: http://www.educause.edu/ir/library/pdf/eqm0137.pdf.

Trott, B. (2001). Audio portals give Web sites the gift of speech. *Infoworld,* February 12. Retrieved April 26, 2002, from: http://www.infoworld.com/articles/hn/xml/01/02/12/010212hnetrend.xml.

Chapter VII

The Next Generation of Internet Portals

Ali Jafari
IUPUI, USA

ABSTRACT

Today's portals bring together existing technologies in useful, innovative ways, but they don't scratch the surface of what is possible. The constant build-up of information and resources on the World Wide Web demands a smarter more advanced portal technology that offers dynamic, personalized, customized, and intelligent services. This chapter discusses next-generation portals and the requirement that they come to know their users and understand their individual interests and preferences. It describes a new generation of portals that have a level of autonomy, making informed, logical decisions and performing useful tasks on behalf of their members. The chapter highlights the role of artificial intelligence in framing the next generation of portal technology and in developing their capabilities for learning about their users.

INTRODUCTION

Today, portal technology is in its infancy. We have just begun to understand and appreciate the dynamic nature of portals and to recognize the need for intelligent user interfaces. The constant build-up of information and resources on the World

Wide Web demands a smarter, more advanced portal technology that offers dynamic, personalized, customized and intelligent services. Next-generation portals must really *know* their members and understand their individual interests and preferences. Furthermore, we need the next generation of portals to have some level of autonomy, making informed, logical decisions and performing useful tasks on behalf of their members. We need to consider the use of artificial intelligence in framing the next generation of portal technology. And finally, we would like future portals to have learning capabilities. The more a member uses the portal, the better the portal should know the member and the member's preferences.

The next generation portal will be able to offer personalized professional services similar to those provided by an experienced administrative assistant or a secretary. For instance, a good administrative assistant knows the kinds of internal and external news that the executive likes to see, and can sort and prioritize that information for the executive's attention e-mail. High priority items might include a new committee being formed that the boss should know about, important social events that he should attend, e-mail messages that he needs to act upon immediately, an important phone message from the vice president, a budget proposal due next month or a call for help from one of his employees who is in trouble and needs his attention.

A human secretary is able to offer these services to the boss because he or she has an extensive knowledge of what the boss likes to know and wants to do. Through repeated interactions with—and feedback from—the boss, the secretary becomes more expert at this. This massive amount of information about the business needs and personal preferences of the boss assists the secretary in acting as an expert agent, filtering the kinds and amount of information the executive needs to perform at optimal efficiency. A trusted secretary also has a certain amount of autonomy to make decisions and perform tasks. For example, the boss may not want to meet anyone on Mondays except in cases with a certain degree of urgency. In filtering the boss's calls, the secretary acts as an intelligent filter and a decision maker on the behalf of the boss.

INTELLIGENT AGENTS

Now let's consider offering some of these services using a series of computer programs within a campus portal environment. For instance, the portal may employ a series of programs called "intelligent agents" to act as a digital secretary. The digital secretary is similar to the human secretary and can offer certain personalized services to its owner. The first time a member signs on to the portal, he or she can access the digital secretary to configure it and other agents by selecting from a menu of personal preferences. For instance, the digital secretary might be configured to prioritize e-mail messages according to a member's interests before listing them in the e-mail channel of the campus portal. Similarly, the portal might sort important campus news that the user should be made aware of, based on the preferences the user has specified, as well as an analysis of the user's past usage habits.

The digital secretary is just one example of the use of intelligent agents to make a portal system smarter. Intelligent agents can be linked to a variety of applications and database software running within a portal environment. Each member of a portal has a personal set of intelligent agents that can be configured to offer personal services. The primary function of an intelligent agent is to help a user better utilize and interact with the portal environment. It is assumed that artificial intelligence (AI) is involved and that a certain degree of autonomous problem-solving ability is present in agent-based technology systems. Nicholas Negroponte (1995) talks about agents as perfect helpers. Another example of an intelligent agent would be a "digital-sister-in-law" that you ask for movie suggestions. Because the agent knows you and your movie preferences and has extensive knowledge about movies and reviews, it can intelligently advise you about what movie to see; it is expert about both movies and you. Table 1 illustrates the roles of intelligent agents in portal applications.

Intelligent agents can offer various suites of services and contain characteristics as discussed below.

Knowledge of Users

Imagine a campus portal that displayed the exact set of news headlines you wanted to read every morning. The news channel would sort the news according to its data about your personal preferences and past usage. For instance, your international news might come from news.bbc.co.uk, your sports news from nba.com, your professional news from the chronicle.com and your campus news

Table 1. Intelligent Agents' Roles in Portal Environments

Functions	Advantages	Benefits
Automation	**Performs repetitive tasks:** Send e-mail to students with overdue assignments.	Increased productivity
Notification	**Informs users' of events of significance:** Inform me about students with two weeks overdue assignments.	Reduced workload
Learning	**Learns users behavior:** Sort the news headlines in my news channel according to my past usage.	Proactive personal assistance
Tutoring	**Coaches users in context:** Offer me additional tutoring materials in my weak math area.	Reduced training

from the faculty side of the campus news service. Your sports news only includes two headlines regarding basketball because basketball is your main sports interest. Your international news mainly includes items related to the United Nations because of your teaching and research interests in international affairs. Your campus news includes only items that faculty members in your field and of your rank would care about. Under this model, you would be less likely to need to visit a variety of Internet news sites. You would get most of the news you wanted when you signed on to the campus portal.

Intelligent portals come to know their members by accessing personal information from members' personal profiles. Such a profile consists of various types of data describing the member's role, interests, preferences and usage history. The more a portal knows a user, the more precise and useful the information and resources it can offer to the portal user. In other words, the intelligence of the portal increases as more personal data is stored in each user's profile. The user profile is a database or a series of databases consisting of many tables, each storing certain types of information about the users.

The portal receives personal information from various channels. Basic user information can be dynamically obtained from the institutional databases. For instance, information can be obtained about the member's department, major, minor, institutional role, etc. A second set of data is personal preferences that are actively selected by each individual member. Examples might include research interests, news interests, entertainment interests, etc. The third set of data is the usage data. Usage data can be collected dynamically by logging members' navigational and usage information. Examples include the news sites the user visits regularly, the types of campus news the user follows (e.g., sport news, faculty development news), etc.

Under this model, we can expect the next generation of portals to contain ever larger amounts of usage data about each member. Databases will play an increasingly more important and active role in the operation of portals. Consequently, we should expect to hear more expressions of concern about the collection and storage of personal data and should anticipate the need for institutional policies on collection, use, security and privacy of personal data.

Subject Matter Expertise

Intelligent portals will offer expert services by drawing upon collections of subject matter expertise. These collections may be built from external, mobile intelligent agents tuned to a given field of knowledge. For instance, a faculty member may be interested in learning more about new research grant opportunities within her field. An intelligent portal can consult with external mobile agents to deliver dynamic reports to the user's portal according to her research interest. The information can then be posted automatically to her personal portal where it receives her attention. In this way, the portal can offer "digital reference librarian" services on any subject. An agent can be specialized for any subject but will have the capability to consult with other agents to provide "the big picture."

Autonomy

Future intelligent portals will have the authority to make certain decisions and perform certain tasks on users' behalf autonomously. For instance, a faculty member's intelligent portal could automatically send e-mail messages to his students about missing the deadline for submitting a class assignment. It could include a reminder that points will be deducted from the student's grade if the assignment is not submitted within an extended deadline. Delegating this responsibility to the intelligent portal saves the instructor time by relieving him of the obligation to visit the course drop box to check off those students who have not completed their assignments.

Trainability

The more an intelligent portal is used to do daily jobs, the more expert the portal becomes about the user. It learns about the user and about the user's interests and preferences by watching the user navigate his or her personal portal. For instance, my own intelligent portal may notice my frequent use of a website link displayed at the bottom of my bookmark channel. The intelligent portal will automatically move that link to the top of the list, giving me more convenient access to a link that I frequently use. If I discontinue using that link, the intelligent portal will automatically move the link down the list, placing new links that I use more frequently at the top. If I did not use a link for a long period, the portal might remove and achieve it in order to clean up my bookmark channel, and showing only those links that I need most often. Similarly, the intelligent portal would put all the sports news at the bottom of my news channel if I demonstrate a consistent lack of interest in reading sports news at a time when my attention is diverted toward major international news.

Another example of trainability is the portal's sorting of new e-mail messages according to a user's usage history. The mail agent that is responsible for organizing my new e-mail messages can monitor and learn about my pattern of reading new messages. The agent notices that I tend to read messages coming from certain individuals and groups before reading the rest of my mail. With this knowledge, the agent can decide how to sort my incoming mail in the way that is most useful to me.

TYPES OF INTELLIGENT AGENTS FOR EDUCATIONAL APPLICATIONS

A large number of intelligent agents can be integrated into a campus portal environment. I have divided the intelligent agents into three major categories or groups. They are: Digital Teaching Assistant (TA), Digital Tutor and Digital Secretary. Each group of agents is conceptualized to perform certain tasks normally carried out by a human being. Each group may consist of one or more intelligent agents focusing on particular tasks within a portal environment. These agents may communicate with their human clients (users of a campus portal) using a combination of text, graphics, speech, digitally rendered facial expressions and voice.

Digital Teaching Assistants

The intelligent agents acting as a Digital Teaching Assistant assist the teacher (instructor or other members of a teaching group) in various teaching functions often performed by human teaching assistants. The Digital TA is a personal agent that may be configured by its owner, the human instructor, at the beginning of a semester. This configuration could include, for example, the agent's level of autonomy to send overdue notices to students on behalf of the instructor, sending statistical grade reports to students with lower class ranking, and the like.

The Digital TA is most useful in distance learning applications. In that learning environment, the instructor is physically isolated from the students, not necessarily knowing if and when they have worked on an assignment, for how long or what types of collaboration were used. Thus, the instructor is incapable of dynamic assessment of the students' work. The teacher is mostly unaware of any given student's progress until an exam is taken, or until the student submits an assignment (or drops out of the course!). In terms of student retention, it is important that the instructor be constantly aware of each student's participation in a course, and assist discouraged students before they drop out. As will be explained, a Digital TA can be invaluable in this role. Additionally, the Digital TA can assist a course instructor

Figure 1. Configuration of an Inactivity Agent

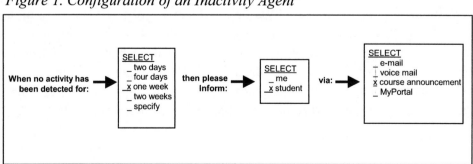

with various course operation and maintenance chores, in much the same way that a human TA helps an instructor in campus-based instructional situations.

Figure 1 suggests a configuration procedure for programming a Digital TA to act as an "inactivity agent." In this example, the agent is configured by the instructor to send e-mail messages to students who show evidence of long periods of inactivity in their courses (e.g., lack of collaboration on the class message board, no record of downloading the reading assignments, not taking quizzes). Note that this is a very simplistic configuration of the agent. In a more advanced example, the agent could be configured to continue monitoring student behavior after the initial notice was sent. This might include sending an additional notice with progressively stronger language if the student fails to respond to a specified number of initial messages. Ultimately, if the student failed to fulfill the Digital TA's parameters for an appropriate increase in effort, the agent would notify the course instructor that the student was at risk. Along with this notification to the course instructor, the agent could provide additional background information about the troubled student, including past submission records, grades, class ranking, etc. (see Figure 2). This abundance of information empowers the instructor to take quick and appropriate action to address the needs of a troubled student.

Figure 2. Sample Display of Students at Risk

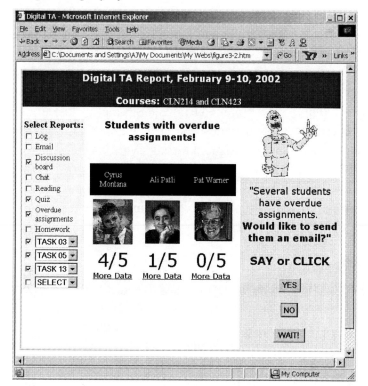

As noted earlier, the Digital TA intelligent agents could include a series of agents. The inactivity agent illustrated here is only one example. The Cheat Buster Intelligent Agent described below is another example of an intelligent agent that could be included in the Digital TA group.

Digital Tutor

Intelligent agents acting as a Digital Tutor assist the learners (students) with specific learning needs, just as a human tutor or classmate might do. The Digital Tutor, for example, could act as a smart search engine, finding specific resources to meet the student's learning needs. Here, too, the intelligent agent is expert not just about a content area, but also about individual students. Depending on the level of its sophistication, the Digital Tutor could become more expert and useful as it provides more assistance to a student and receives more feedback, both from the student's online behavior and through assessment tools integrated into the learning environment. For instance, consider an online distance-learning course in which a student has difficulties understanding new subject material. The Digital Tutor would have access to outside mobile agents that could help it search for and identify appropriate pedagogic resources, fine-tuned by information drawn from the student's profile. Accessing student profiles and learning from them students' strengths and weaknesses empowers the Digital Tutor to apply useful resources to the students' learning objectives.

Student profiles used by a Digital Tutor would consist of data dynamically obtained from a Student Information System (SIS), information entered by the student into his or her own profile, or usage data dynamically obtained from the course management access log file, including both present and past courses. Examples of dynamic data obtained from the SIS include the student's major, previously taken courses and grades. A smarter Digital Tutor might utilize data based on a student's past assessment through various learning modules. For example, the Digital Tutor might find that a student taking a second-year college English course did very poorly in the grammar part of his first-year course. Based on this information, the Digital Tutor could suggest that the student complete supplemental grammar exercises.

Digital Secretary

The introduction to this chapter introduced the idea of a Digital Secretary. The intelligent agents acting in that role assist students and instructors by performing various logistical and administrative functions. Like a human secretary, the Digital Secretary performs tasks as mandated and directed by its supervisor—in this case, the human being at the keyboard. An example of a Digital Secretary is the "out of office" e-mail notification service offered by the Microsoft Outlook messaging software. The user may program Outlook to send an automatic e-mail notification in response to messages received during a specific time period. The Digital Secretary conceptualized in this chapter, however, offers much more intelligent and sophisticated services. For instance, consider the case of an instructor who would like to send different auto-response e-mails to students taking a particular section of his undergraduate course than to his other correspondents.

The user and environmental profile settings may provide the initial configuration of the Digital Secretary agent. Scheduling a meeting, finding a colleague with similar research interests available over weekends or finding the best math students who might serve as mentors are examples of tasks undertaken by a Digital Secretary within a teaching and learning environment. A Digital Secretary could also be used by other constituents of an educational institution who are not directly involved in teaching and learning (e.g., administrative staff, alumni and parents).

HOW DO INTELLIGENT AGENTS ACT IN TEACHING AND LEARNING SITUATIONS?

Intelligent agents such as the Digital TA, Tutor and Secretary operate within a portal environment. Each portal member (student, teacher and staff) has access to a selection of personal intelligent agents. Each agent can be configured or programmed by the member. For instance, a teacher can program his agent to send e-mail notification to students with grades lower than C who additionally did not participate in collaboration activities hosted on the classroom message forum.

As illustrated in Figure 1, the agent can be programmed to sequentially monitor certain incidents, compare them with preset trigger and perform certain tasks on behalf of the owner. Depending on the type of agent, access for configuration of the agents could be provided through the "MyPortal" section of a campus portal. The agents could be course-specific or multipurpose (i.e., an agent that monitors certain activities in a specific course versus one that monitors all the courses that a student is taking or an instructor is teaching).

Figure 3 illustrates the basic architecture of intelligent agents for teaching and learning. As shown, an agent may have access to a variety of dynamic and static data, including data obtained from the campus Student Information System, course management system and student profile databases. Based on this information and on configuration settings provided by the owner of an agent or the course and system administrator, the agent will be able to "think" and perform intelligent actions. Because artificial intelligence requires a massive amount of data processing power, it might be necessary to run the agent software on dedicated computer servers. Furthermore, it might be necessary or at least beneficial to distribute the various tasks performed by an agent among several computer servers.

Figure 3. Basic Operation of Intelligent Agents

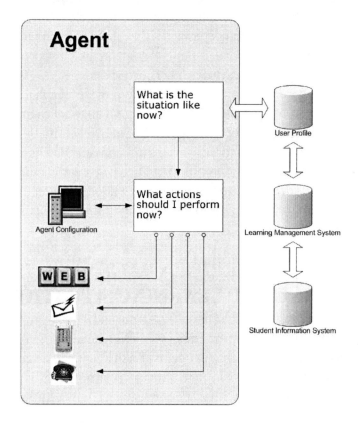

SCENARIOS FROM AN AGENT-BASED TEACHING AND LEARNING ENVIRONMENT

This chapter is intended to define and conceptualize applications of intelligent agents in portal environments, especially for campus portals. One common method of painting the big picture in defining new applications is by presenting them in a realistic scenario format. The following section presents a scenario highlighting the functionality of Digital TA Intelligent Agents.

... Monday morning, Professor Amy Warner of Indiana University Purdue University Indianapolis arrives at her office around 9:00 a.m. She is teaching two courses that meet twice a week and are complemented by course management software within the campus portal environment. Professor Warner switches on her computer, logs into her campus portal, clicks on the Digital TA icon and begins organizing papers on her desk. A Digital TA named Angie appears on her desktop as an animated character in the top right corner of the computer screen. "Good morning, Professor Warner, here are the activities of your CNL214 and CLN423 courses. Over the past weekend there were moderate activities on your CLN214 website."

Professor Warner, while browsing through the campus newspaper, looks up at the computer monitor on the other side of her desk. An XY graph in the middle of the screen shows the students' activities plotted against date and time over the last three days. "Angie, next," Professor Warner says to see the next activity report.

"There are three students with overdue assignments," Angie says. "Would you like me to send them your generic 'overdue assignment' e-mail notice?" Professor Warner, holding coffee in one hand and sorting books with other hand, looks up and recognizes the pictures of the three students on the computer screen (see Figure 2). She asks the Digital TA to send them the generic e-mail message, reminding them about their overdue assignment and the deduction of ten points that will result if they do not submit their assignment within the next two days. The Digital TA agent automatically sends these messages to the students and records the action in the course grade book. Professor Warner also notices that Cyrus Montana has missed the deadlines for four out of five assignments during the course of this semester. She makes a mental note that she may want to talk to Cyrus after her lecture on Friday.

On Tuesday, Professor Warner gives an online quiz to her CLN214 class. On Wednesday morning, after she logs on to her campus portal, she receives an alert from the Cheat Buster Intelligent Agent. "Excuse me, Professor Warner, I have detected many similar quiz answers between Cyrus Montana's and Pat Warner's tests." The agent continues, "There is 92% similarity between right and wrong answers in their last quiz, and both took the quiz at the same time from two adjacent computer workstations in the University Library."

The Cheat Buster Intelligent Agent then displays a three-dimensional graphic on the screen, highlighting similar answers. It also provides statistical analysis of the past six quizzes, highlighting other similarities between the two students' results. Professor Warner sends e-mail messages to Cyrus and Pat asking them to meet with her after class on Friday. ...

NEW INTERFACES

The next generation of portals will offer new types of communication channels in addition to the current common Web browser interface being used by personal computers. Some commercial portal systems are currently offering limited services via Personal Digital Assistants (PDAs). The use of PDA technology as a complementary interface to the Web browser will substantially increase in the near future. The widespread use of cell (mobile) phones, for instance, will create interest among the cell phone users to access their Internet portals via their phones when they are away from their desktop computers. The difficult part of offering campus portal services via cell phones and PDAs is the difficult task of redesigning the user interfaces. It is not an easy task to repackage the massive amount of information presented on a high-resolution Web browser into the small display area of a cell phone or PDA screen. Additionally, cell phones and PDAs offer limited input capabilities. The ergonomics benefits of a full-size computer keyboard and mouse cannot be offered within the small physical size of cell phone and PDA packages.

REFERENCE

Negroponte, N. (1995). *Being Digital*. London: Hodder & Stoughton.

SECTION II

CASE STUDIES OF CAMPUS PORTALS

Chapter VIII

Indiana University's Enterprise Portal as a Service Delivery Framework

James Thomas
Indiana University, USA

ABSTRACT

Indiana University has embarked on a journey to create an enterprise portal, a place to which faculty, staff, students, alumni, prospective students and others will travel to uncover a broad array of dynamic Web services. "OneStart" (http://onestart.iu.edu) provides a compelling place for faculty, staff and students to find services with the intent of developing life-long campus citizens of IU. Indiana University is a large and complex institution consisting of eight campuses, over 92,000 students and 445,000 alumni. Developing one enterprise portal for the entire institution presents tremendous challenges. IU's portal strategy hinges on the ability to provide a "service delivery framework." This framework provides single sign-on, customization, personalization and workflow. By making use of component based design methodologies, the framework also provides a flexible publishing environment allowing services to be created by different Web development teams across all IU campuses. This chapter describes the motivation and strategy behind the creation of OneStart, the enterprise portal service delivery framework at Indiana University.

INTRODUCTION

Indiana University has embarked on a journey to create the next generation Web portal, a place to which faculty, staff, students, alumni, prospective students and others will travel to uncover a broad array of dynamic Web services. Many of these services will be available at anytime and from anywhere. Indiana University's enterprise portal, "OneStart" (http://onestart.iu.edu), provides a compelling place for faculty, staff and students to travel and explore with the intent of developing life-long campus citizens of Indiana University.

Why build an enterprise portal? Today it seems that every institution in higher education either has an enterprise portal or is talking of building one. Many people are certain that their campus needs an enterprise portal. They may or may not be sure why. The reasons seem so obvious to everyone else that it doesn't seem important to ask. All of the other universities are building enterprise portals, so many feel that they must get started right away on building one of their own.

Starting a portal project for these reasons alone is a mistake. Any successful portal project team begins by asking itself what are the business problems on campus that it is trying to solve? This may seem an obvious beginning, but it is frequently overlooked. Technology is a great problem solver. Enterprise portals are one of the latest technological innovations. What could possibly be wrong with having one? The fact is that there may be more fundamental problems at an institution that will only be further exposed by an enterprise portal project. While portals appear to be here to stay, it is critically important that the team have a comprehensive understanding of the problems that need solutions. For whom are they building the portal? How will it help the users? A successful portal project begins with the answers to these questions. For Indiana University, the answers led down the path to building an enterprise portal.

Indiana University is a large and complex institution consisting of eight campuses, each with its own objectives and culture. IU has over 92,000 students and 445,000 alumni worldwide. The number of online services available across the institution is enormous and continues to grow at a rapid pace. An attempt to develop *one* enterprise portal for the entire institution presents tremendous challenges. In order to make the best use of scarce technology resources, the approach IU took hinged on the ability to provide a "service delivery framework." This framework needed to provide all of the features of a portal such as single sign-on, customization, personalization and workflow. However, the framework itself needed to be built using components with standardized interoperability features. This strategy allows the services themselves, the real content for the portal, to be created by different Web development teams across all campuses using a wide variety of Web development toolsets.

Enterprise portals are different at every institution. An enterprise portal's characteristics and features are dependent upon the problems the institution is trying to solve and its vision for the future. Each set of problems and requirements is unique to the institution. This chapter is an attempt to describe the motivation and strategy behind the creation of OneStart, the enterprise portal service delivery framework at Indiana University. This chapter offers some insight into ways of approaching enterprise portal projects and generating ideas for others who are beginning or tuning their own projects.

VISION FOR ONESTART—INDIANA UNIVERSITY'S ENTERPRISE PORTAL

The seeds for Indiana University's enterprise portal project were sown with the publication of the information technology strategic plan, "Architecture for the 21st Century: An Information Technology Strategic Plan for Indiana University" (McRobbie, 1998). One of the problems identified in this document includes the broadening base of information consumers. The use of Information Technology (IT) has become commonplace among students, faculty and staff at higher education institutions. IT is no longer isolated to complex, back-office administrative systems used to automate complex business processes. The complex, monolithic, "stovepiped" application systems of the past are now obsolete. IT consumers today use numerous application systems delivering a broad range of services. The current broad and diverse base of users requires a different approach to developing applications.

Aging and disparate legacy information systems that lack flexibility and integration no longer meet the demands of these constituents. Students, faculty and staff want more convenient access to, and more efficient delivery of administrative and academic services. In this age of disintermediation, people do more and more things for themselves. ATM machines, online shopping and electronic forms bring valuable services directly to people conveniently and efficiently. Expectations are that universities will keep up with these trends. Fundamental transactions like registering for classes, paying a tuition bill or administering grades should not require shuffling paper documents and wearing down shoe leather.

Institutions of higher education are also dealing with larger quantities of "non-traditional" students. Baby boomers beginning second careers, employed professionals learning new skills and retirees taking up new hobbies have changed our campus communities. Dealing with these individuals in traditional ways invites failure. We must strive to make access to services more convenient and effective for them. At the same time, we must also work to provide opportunity for these individuals to feel a part of our campus communities, even if they do not physically step onto campus. A fundamental requirement of any campus portal is the ability to create a virtual community for all campus citizens. The enterprise portal should not only provide access to a broad set of services but should also provide a compelling place for its citizens to "gather" to discuss class projects, current events or simply socialize. The enterprise portal can be a tool that allows the non-traditional student to take part in the culture of the campus.

Having a strategic plan to address the fundamental problems at your institution will lay a solid foundation for all future IT development. At Indiana University, the initiatives outlined in our strategic plan provided an unprecedented opportunity to begin the seamless integration of enterprise application systems and present them through a coordinated, unified front end—an enterprise "portal" to the enterprise systems and the services they offer. The ultimate goal of the OneStart project is to provide a higher level of service to the faculty, staff and students of Indiana University in the form of an easy-to-use, convenient and integrated delivery framework for all of the online services available on our campuses.

Indiana University's strategic plan identifies 68 individual action items designed to allow Indiana University to "take the next step in institutional academic excellence and move into the very top tier of the nation's public universities." OneStart addresses three of these action items specifically—Common Interface, Architecture and User-Centered Design. The specific entries from the strategic plan read as follows:

Common Interface

Users can be categorized into three groups. There are back-office users in the operational units doing transaction processing on a daily basis and understanding more of the complexities of the systems and data. There are students, faculty and staff who need only occasional, casual access to the information. And, somewhere in between these two groups, are the service providers in various schools, departments and programs who need to use a diverse set of information systems. These last two groups need and will benefit most from consistent and integrated access to the applications, data and systems.

ACTION 37: UITS (University Information Technology Services), working with the users of IU's administrative systems, should develop a common interface environment that will support the efficient and effective accomplishment of the day-to-day administrative tasks of the university.

This common interface environment should be implemented across all commonly used desktop computing platforms and operating systems.

Architecture

Technical and information architecture, technology standards and enterprise business plans are critical to long-term success and stability of information systems. These ensure that central and distributed computing systems can operate together and that disparate data sources can be combined, analyzed and reported. Future developments or acquisitions will require a multi-tiered architecture that supports the development of components as a requirement for facilitating change and performance. In this architecture the presentation, business logic and data are separate entities. This facilitates business, data and technology changes while minimizing the impact and cost to the institution. There should also be greater investment in network application products (e.g., Winframe and Hydra), to support the demands for existing, large workstation-based applications. A central server would execute these applications and enable the use of lower performance workstations, avoid problems of application distribution and minimize network bandwidth demands.

ACTION 38: UITS should enhance its current information and IT architectures to include the use of "thin client" technologies, and employ multi-tiered architectures in future software development.

IU's information systems architecture will depend, too, on development and support of a production-quality UNIX computing environment and on the use of software component technologies for systems development.

Usability Laboratory/User-Centered Design

Usability of information systems from the users' perspective in meeting their requirements is key to their success. A goal for all of the university's information systems is that technologies should be selected or developed which are appropriate to the needs of their various users and suitable to the business need that is being addressed. To achieve this end, there should be a commitment made to user-centered design, employing usability studies and bringing an explicit focus on end-user needs and requirements to the design of all university information systems.

ACTION 44: UITS should incorporate user-centered design techniques and Usability Lab testing into all major systems development projects.

To assist in this effort, the UIS Division has built up expertise in usability studies and user-centered design, and has developed and equipped a modern Usability Laboratory.

Action 37, the "Common Interface" initiative, speaks directly to the implementation of a portal, a common interface for administrative tasks. A portal provides a single interface for faculty, staff, students and service providers to use for day-to-day activities. An enterprise portal initiative takes this concept a bit further by including access to other valuable services, such as information, research and collaboration resources, all in one environment with a common user interface.

Action 38, the "Architecture" initiative, identifies multi-tier and thin-client platforms as the strategic direction of the institution. This initiative alludes to another fundamental development strategy being adopted within the enterprise portal framework—a component-based design approach to the delivery of applications. A component-based approach allows for the reuse of code and, perhaps more importantly, the adaptability of that code to changes required by future needs.

With this approach to development, the user interface components are separate from the business logic components that are separate from the data access components. These components have published application programming interfaces (APIs) that allow for reuse by other applications, allowing development teams to avoid reinventing the proverbial wheel. Emerging standards within the world of Web services promise to make the component interface independent of a component's implementation. Developers can simply plug their services into the portal irrespective of the development tools they used to build those services.

The component-based design approach also provides flexibility to change in the future. It is impossible to predict exactly what technological advances will be made in the next three to five years. The only certainty is that technology will change. Therefore, we must build our applications in such a way that they can easily adapt. This approach will extend the life of our enterprise portal.

Action 44, the "User-Centered Design" initiative, is what portals are really all about. Our enterprise development model has shifted from developing applications for facilitating back-office business processes to facilitating the user's ability to access services with ease and convenience. The users are now in control. Portals must be designed and developed with continuous input from the users. The faculty, staff and students of Indiana University tell us how our enterprise portal should work and what services should be available there.

In the past, a development team might meet with the users of a system early in the requirements gathering and design phase of the project. After a few months, the developers would appear with a demo of a completed application that might or might not have been what the users expected. In some cases, the users' needs might have changed since they last met with the development team. This model is flawed.

With a user-centered approach, the functional users of the system are a part of the development team from the very beginning until implementation day and beyond. The users are the experts and understand exactly what services they need out of a given application. It is only through a constant opportunity for input and feedback that system developers can make the right decisions and keep pace with change. Functional users are a part of the development team, and the application must be developed in iterations providing ample opportunity for their input.

Beyond user participation throughout the development process, ongoing usability and accessibility testing provides validation and feedback on development progress. With the broadening numbers of information consumers also comes a larger variety of computing savvy and expertise. Self-service type applications demand better content organization, logical process flow and standard user interfaces. Frequent and ongoing usability testing can insure that the application is easy to use and a satisfying experience to the end-user.

With recent federal legislation requiring adherence to standard accessibility guidelines for public institutions, state-sponsored universities like Indiana University must also insure that their information systems are accessible to the disabled. Indiana University is home to the award-winning Center for Adaptive Technologies, which assists information systems development teams in their efforts to provide accessible sites. The OneStart project team is dedicated to providing equal access for the disabled to the services provided within the portal service delivery framework.

These three initiatives outlined in the strategic plan laid a foundation for the beginnings of an enterprise portal at Indiana University. In March 2000, nearly two years into the strategic plan, a specific plan for an enterprise portal was created. A two-day Joint Application Development (JAD) session was conducted with several senior technical staff and their functional counterparts in various administrative offices including Human Resources, Financial Management, Accounts Payable, Purchasing, Registrar and Bursar. The purpose of this JAD session was to define and identify the required deliverables in administrative computing for the next three to five years. After two solid days of brainstorming and prioritization of goals, some of the most important deliverables identified included single sign-on authentication, role-based access, personalized desktops, dynamic routing and approval, and a standard set of business rules. In addition, the group suggested that any applications developed in the new environment would be designed with a standard set of guiding principles to facilitate 24 X/7 availability (when business needs require it), reusable components, usability and accessibility, imaging/document management and directory services. After the JAD, the basic requirements of the portal service delivery framework had begun to take shape.

The JAD session included some strategy discussions regarding the design of future applications. The discussion identified three separate design "layers" that would make up the enterprise portal framework: an infrastructure layer, a service layer and a user interface layer.

The infrastructure layer, or EDEN (Enterprise Development ENvironment), consists of component-based, reusable standard modules upon which services can be built. EDEN components are fundamental building blocks with published interfaces that are independent of their implementation. These infrastructure components, among which are many of the building blocks commonly known as "middleware," allow the portal to provide a consistent interface to a disparate set of back-ends including Enterprise Resource Planning systems (PeopleSoft in the case of IU), mainframe (CICS Cobol) and numerous departmental legacy systems. One major deliverable for IU's EDEN infrastructure is the integrated workflow routing engine for the routing and approval of electronic documents.

Using a component-based approach, EDEN provides a flexible, scalable and extensible environment allowing OneStart to be agile in the fast-paced and unpredictable worlds of information technology and higher education. An infrastructure layer like EDEN is the foundation for any successful enterprise portal project. The majority of the major technical issues contained within an enterprise portal initiative are with integration at the middleware layer. Without all of the hard work that takes place in the middleware space, the ability to meet user demands will be limited.

The service layer builds upon EDEN by taking advantage of pre-written standardized interfaces and component-based services that exploit core institutional data. The portal services, or "channels," in the service layer consist of services provided by application development teams across all IU departments and campuses. Application developers can reuse and combine EDEN components to quickly create new and enhanced Web services. Diverse services such as tuition payment, theater ticket purchase and class registration can be developed by different people. This requires a new approach to delivering applications. These individual services are no longer self-contained within large, complex, monolithic applications. Each application developer must now think of his or her system as a set of discrete services that are delivered to users via the portal framework. These most popular features within a given application become channels, windows of Web content, each displayed in its own subsection of the portal page.

Finally, the user interface layer consists of the portal framework components that provide the portal presentation. The portal interface is made up of default portal pages whose selection is based on a user's role, the channel layouts that make up those pages, role-based filtering of portal content, personalization features, and the general look and feel of the portal. If OneStart is to become a place where people do their business every day, it must present a very compelling and comfortable environment within which the user may work. If OneStart is to provide a virtual campus community for users to visit, the user interface must encourage return visits. Therefore, the importance of the user interface components is paramount. If the users do not enjoy the experience or find it easy to use, the portal will not fulfill its goals.

With the strategic plan initiatives and the JAD session results, IU's vision for an enterprise portal took shape. OneStart is about providing better service to the faculty, staff and students. The portal pulls the best services from all applications, websites and collaboration tools into one universal environment. Users can easily navigate through this environment to access their favorite services. The service delivery framework provides flexibility allowing adaptation to each individual's requirements. The framework has been built to allow distributed development of portal services by all campus development teams while allowing for significant changes and adaptation in the future.

The first generation of the Web was about sharing information. The Web is an ideal way to make static information content available globally. An emerging trend is the delivery of applications via the Web. These "next generation" Web services provide the ability to conduct business transactions via the Web. OneStart is a "next generation" portal that includes access to Web-based services that provide access not only to information but also to self-service applications and collaboration tools. Students can register for classes, check their grades, apply for admission, chat with classmates and reserve a library book all in one place. A faculty member can advise a student, access research tools and collaborate with colleagues. Staff members can improve their efficiency on the job by creating desktops comprising those services they use most often throughout the day. Staff also benefit in that they can make better decisions given their enhanced access to information and tools for collaboration.

Finally, it is important to understand that OneStart is not an information portal. Many campus homepages are limited to that; they provide access to just about any information available about a particular institution, but they do little to integrate or personalize the services the campus enterprise offers. Informational homepages provide a great service. They are the first place people who are unfamiliar with your institution will go to find out more about it. But Indiana University is not seeking to replace its homepage with OneStart. OneStart is truly a "next generation" portal.

DEVELOPING ONESTART—BUILD OR BUY?

Once IU had clearly defined its portal vision, it was time to get started. Implementation of the concepts and ideas behind the OneStart portal required the support of campus administration. A project of this magnitude requires cooperation and collaboration across many different groups throughout the institution. Support at the executive level increases the likelihood that the necessary level of cooperation can be attained. The OneStart vision was communicated as clearly and broadly as possible during the summer of 2000. Early in the process, OneStart project team members engaged IU students, faculty and staff seeking ideas and suggestions for making the portal a success. In general, the response was overwhelmingly positive. Most people saw the value in an enterprise portal solution and were eager to have

> ## YOUR DECISION TO BUILD OR BUY
>
> The build vs. buy decision is one that must be made on a case-by-case basis depending on a number of factors and characteristics of a given institution. First the institution must evaluate its resources and time requirements. Do they have the necessary resources to allocate to a portal-building initiative? Depending on the project scope, most portal initiatives require a minimum of two-to-three dedicated full-time developers; many projects have required 10 or more! If an institution is not prepared to commit resources at this level, building a portal may not be an option.
>
> It is also important to examine integration and deployment issues. What are the institution's preferred or standard development environments, especially for those systems requiring integration? What types of environments is the institution prepared to support? It is critical to understand all of the requirements for an enterprise portal solution. Each institution must identify potential constraints or issues with specific solutions or technologies.
>
> Once an institution has analyzed its own requirements, resources and issues, it can begin entertaining vendors. First, the institution must develop an objective list of evaluation criteria by which to compare and contrast solutions. Then the list must be prioritized. What items are most important to the portal initiative? This will make working with vendors easier. Discussions will be more direct and to the point. Finally, the institution should implement a decision-making process that includes all stakeholders in the portal initiative. All participants need to feel they contributed and played a role in determining the plans for the enterprise portal. At Indiana, we mapped out three separate plans that reflected the costs and timelines associated with each option—buy, buy and build, and build. Then by simply comparing the plans using our list of criteria and understanding our resource constraints, we were able to make an informed decision.
>
> If an institution chooses to buy a product, it should be sure to choose a vendor or partner that will be around for at least two-to-five years.
>
> An enterprise portal is one of the most important applications at an institution. Whichever option is chosen—buy, build or a combination of the two—an institution should be sure it is comfortable with the amount of control it will have over the future of the portal.

the service available. So, support was gradually built. Communicating plans for OneStart as broadly as possible enabled everyone to begin thinking about OneStart as the service delivery framework of the future.

After laying the groundwork to get the necessary support, it was time to come up with a plan for implementing such an ambitious project. The first major question that needed to be answered in order to create a plan was whether the enterprise portal framework should be bought, built or some combination of the two. An analysis of the leading portal vendors and their offerings ensued. Some of the leading portal vendors in the late summer of 2000 were Vignette, Tibco, Plumtree, Viador, Sequoia, Epicentric and Blue Martini. Many of these vendors offered what were called "complete" enterprise portal solutions.

In general, with the economy still in pretty good shape, the pricing models for these products were quite expensive. Many were based on a "per user" pricing model. Given that Indiana University has over 96,000 students, any pricing model based on the number of users resulted in extremely high quotes. So, the "buy" option did not seem promising.

While quotes ranged anywhere from $500,000 to $3,000,000 for the software, these solutions would still have required a large amount of integration and configuration work unique to each institution. This meant additional resources and time would have to be devoted to the project. Including the additional staffing, hardware, integration and maintenance costs for such a solution, IU estimated that it would require between $3 and $7 million to build the portal it envisioned. Those prices were simply out of reach.

Pricing was only part of the problem. Another problem was the immaturity of the leaders in the portal market. They were mostly new startup companies with very unpredictable futures. Given the risks associated with these vendors and the dollar amounts being discussed, it was not prudent at that time for IU to attempt to buy and implement a commercial portal solution.

At that time, there were also portal frameworks available. Microsoft, Oracle and IBM offered portal-building tools designed to make it easier and faster to build and deploy an enterprise portal. Their solutions offered a "buy and build" option. At the time, IU judged that these tools were still immature. They did not provide functionality to do many of the things IU was trying to do. The tools were in early releases and still had many problems that needed to be resolved. The amount of time required to buy, build, and configure an enterprise portal with one of these solutions was prohibitive. IU's estimates for buying a framework and building a portal meeting our requirements ranged from $2 to $3 million. While considerably less than buying one of the leading "complete" solutions, this was still a large amount of money. Another problem was that some of the tools did not include completely "open" architectures. Indiana University required an "open" approach to allow for integration of legacy and third-party applications while allowing the use of the diverse set of development technologies that often exists within an institution of higher education.

While the framework perspective was a less expensive alternative, it was not without its problems. Many of these companies were established vendors, but were new to the portal market. Portals were not their main source of income. So, IU wondered how committed they would be to supporting their portal products. Adopting one of these solutions seemed a risky proposition, given that the vendors were so new to portals.

Despite the "buy and build" estimates being lower than the buy option, they were still significant numbers due to the additional requirements they imposed for development, configuration and integration with existing systems. In October 2000, the project team was gratified to discover that a report published by the Gartner Group supported our findings regarding the portal options then available. Gartner agreed that the market was risky and unstable at that time.

In the end, the only feasible option for Indiana University was to build its own portal framework. By delivering a portal framework using an open architecture approach, IU could provide a service delivery vehicle that allowed the integration and merging of services from many heterogeneous applications. IU is also banking on the promise of the emerging standards for Web Services (SOAP, WDSL, UDDI, etc.), which eventually will allow an entire world of Web components to be published within a portal framework. Indiana University is in control of its portal, as well as the delivered features and system architecture. Much of the integration work needed to be done with any of the buy or build options. IU reasoned that if they could devote sufficient resources to a portal effort, building it in house was the most logical and fiscally responsible choice.

Building OneStart—A Service Delivery Approach

From the outset of the portal framework building project at IU, it was clear that the resources available for the effort were limited. Any attempt to provide a portal framework *and* develop all of the available services by a single development team would have created a project so large and complex that progress would have been sluggish at best, assuming the project could even have gotten off the ground. IU decided to take a different tack.

The project planners decided to develop a service delivery framework that would allow the convenient and effective delivery of services to faculty, staff and students. The fundamental requirements for the framework consisted of five deliverables—single sign-on authentication, role-based customization, user-centered personalization, integrated workflow and adaptive profiles. The plan was to develop the framework in a manner that made it independent of the individual Web services being delivered. This would allow IU to incorporate all of the Web development work that had already been done into the portal while taking advantage of ongoing work for services to be delivered in the future. As it has evolved, OneStart simply provides a framework that makes it easy for constituents to access the services that they need. This framework also allows service providers to deliver Web applications more efficiently. The developers no longer have to develop their own authentication mechanisms and workflow processes; they can simply "plug in" to these services within the portal framework using standard interfaces. This leaves them with more time to focus on delivering more robust solutions for their respective clientele.

Copyright © 2003, Idea Group Inc. Copying or distributing in print or electronic forms without written permission of Idea Group Inc. is prohibited.

Services delivered via the portal framework provide access to information, self-service applications and collaboration tools. The framework facilitates delivery of all three in the belief that the users will take advantage of these services simultaneously throughout the day in performing their work, research, and business at Indiana University. By delivering a framework that is independent of the various Web development platforms being used, IU has begun to coordinate the broad range of services being offered across all of its campuses into one universal effort. The project team believes that this model and approach are the quickest and most effective way of solving some of the most difficult problems at the institution.

Building OneStart—Methodology

In order to build this "next generation" portal, the OneStart project team began by reviewing existing enterprise application development methodologies. Given the new emphasis on developing a service delivery framework using component-based design principles, older methodologies no longer applied. The goal was to avoid developing stand-alone, monolithic applications for automating back-office administrative processes. This project required a new approach to applications development that incorporated component-based design principles, specifically the separation of the user interface, business logic and data access functionality.

Indiana University needed a methodology that would support the vision of OneStart. To that end, a "living" methodology document was created based upon the experiences of developers and the help of consultants experienced in component-based design. The methodology document continues to be revised as future iterations of the portal are developed. Some of the fundamental components of the methodology include defining business processes and functions, mapping requirements to business objects and usage scenarios, and mapping business objects to components. This process requires that developers focus on the business requirements. Implementation details absolutely should NOT be discussed during this process. This can be extremely taxing for developers who often want to think right away about how to begin coding. The temptation to do so must be resisted.

Many challenges face organizations attempting to implement systems using this type of methodology. First of all, an organization must have well-defined business processes. This is not always the case. It can also be difficult to gain access to valuable functional staff who are capable of clearly explaining the business processes to IT staff. Training of IT staff may be required in some cases. Communication and cooperation among development teams is crucial. In order for the methodology to work, standards for developing, maintaining and publishing services within the framework must be established.

A Web services or component-based approach has several fundamental advantages. First, developers have a large repository of reusable business functions that they can use to rapidly build Web services. Common services no longer need to be replicated in different applications across our campuses. Developers need only worry about the new functionality that their services will provide. This allows them to focus on what is important. Utilizing a shared service that has already been written, debugged, tested and implemented simply requires understanding how to use the published interface. More importantly, if the emerging Web services standards catch on, a whole world of available services will be made available to institutions that have positioned themselves to take advantage of these standards and technologies.

Perhaps an even more important benefit of the component-based approach is that it allows developers to replace specific business functions without affecting the entire application. By abstracting out specific components of an application and deploying them as independent entities, it becomes much easier to enhance existing applications. The future cannot be predicted. Our intent is to build a portal that can easily be adapted, enhanced and modified as new technologies or requirements are discovered. This strategy also provides a "buffer" for any wrong decisions that may be made. It becomes much easier to swap out a particular component or hardware/software solution that may not be meeting the demands or needs of an application. A flexible and open architecture gives OneStart the agility required to extend its life well into the future.

Building OneStart—User-Centered Design

Building a completely user-centered environment is critical for a successful portal. An enterprise portal is intended to make it easier for users to access valuable services. If the portal is difficult to navigate, understand or use, then it will not succeed. A successful portal is a compelling place for the user to visit, one where they feel comfortable navigating and accessing services. Only easy-to-use portals will be adopted by users. The very intent of portals is to simplify access. If a portal is not easy to use, then it is not accomplishing its goals. Information technology professionals often underestimate the importance of usability. It is often thought about only a few weeks prior to going live. This is much too late to have much of a positive impact on the overall usability of the application. Usable design, layout and interface must be considered at the outset of every development initiative.

Indiana University used three fundamental strategies to develop OneStart as a user-centered application. First, potential portal users were involved in the initial design of the application. The university began this process by having test subjects in its usability lab perform a series of tasks on existing portals like Yahoo, MyFidelity and MyExcite. From the results of these tests, the project team developed recommendations for what works and doesn't work on a portal site. Later the team performed more detailed testing with a prototype portal that they had developed based on the initial recommendations from the original portal study. Later releases of the portal, including the first production release, included changes based on recommendations that emerged from the prototype study. This process is ongoing; usability testing continues to be performed upon each release of OneStart. With each study, the project team finds new ways to improve the interface.

The second strategy involves providing opportunities for direct feedback from OneStart users. First, the project team formed role-based advisory groups that were specifically designed for gathering focused input from individuals in the faculty, staff, and student roles at each of the IU campuses. The groups provided input concerning the specific functionality that they would like to see in the portal. The project team also uses these opportunities to get feedback about the users' daily interactions with OneStart. What kinds of problems are they experiencing? What kinds of questions are they asking? Input from these groups is gathered in a very informal setting. In this way, the participants become active members of the OneStart project team. They witness the real impact of their work in each release.

The third usability strategy focuses on accessibility. Indiana University is a state institution and is therefore subject to the accessibility guidelines outlined in Section 508 of the Federal IT Accessibility Initiative. Developing an accessible website presents a number of challenges for developers. The OneStart team began by adhering to the W3C Priority 1 standards outlined at www.w3.org/TR/WAI-WEBCONTENT/. A software product called "Bobby" (http://www.cast.org/bobby) allowed testing of Web applications for accessibility compliance. The OneStart team also worked with the award-winning Adaptive Technology Center at Indiana University to test the portal with users who have visual and mobility impairments. The ATC utilizes sophisticated technology such as Braille readers and voice recognition software. One of the most interesting discoveries from this effort was that the subjects of our study did not want a Web experience different from that of other portal users. They simply wanted access to the same features and tools. The OneStart team continues to make improvements in the accessibility of the application.

Building OneStart—Portal Governance

Indiana University began development of OneStart based upon the shared vision of all portal stakeholders. An executive steering committee was formed to help prioritize the requirements and content of the portal. As requests for new services and functionality in OneStart pile up, it is the charge of this group to determine the project priorities. They determine which tasks get worked on first. This committee is made up of senior administrators from key departments on all IU campuses. As of this writing, the OneStart team is seeking to identify additional members of the steering committee to represent more directly the interests of students, faculty and alumni. With the addition of members for these roles, the portal project is effectively "owned" by the group.

In addition to the executive steering committee, the university's faculty, staff and student advisory groups continue to provide ideas and requirements for future portal releases. The advisory groups focus on ideas for new services and general usability concerns. Any feedback from them requiring additional work is brought before the executive steering committee for prioritization. OneStart is being developed in iterations. Based on user feedback, usability studies and decisions by the steering committee, the deliverables for each release are identified. OneStart can continuously grow and evolve as user requirements and technological advancements change the way services are delivered via the portal. IU expects that the work of these committees will continue throughout the life of the portal itself.

Finally, it is extremely important to determine the services that will appear on the default pages, those pages that appear first upon logging into the OneStart environment. These default pages are based on the users' roles or combinations of roles at the institution. In order to develop complete default pages, the OneStart project team formed focus groups made up of individuals representing each role on each campus. Thirty-two focus groups now determine the content of the default pages for faculty, staff, students and alumni for each of the eight campuses of Indiana University. Evidence suggests that many portal users do not take advantage of personalization features. Most users, estimated to be anywhere from 70% to 85%, will simply use the defaults that they are given upon logging into the portal environment for the first time. Therefore, it is very important that the appropriate services get established on the default pages. This also creates some healthy competition for real estate and placement on the default portal pages. It is up to the focus groups on each campus to decide the content and layout of the default pages for their roles. The default pages can evolve with future releases of the portal to include new and improved services with each release.

Building OneStart—Portal Navigation

Developing the portal navigation strategy proved to be a real challenge. Within portal pages, the windows of sub-content, called channels, consist of links or subsets of Web content that appear in a specific area of a portal page. The existing portals that the project team observed in its early usability studies and prototyping featured two basic styles of navigation. The first approach, referred to as basic navigation, simply replaces the entire portal page, with the user-selected channel content taking up the full screen. This approach works well for pages that do not require much additional navigation beyond the original page. However, if the users subsequently click deeper and deeper into the content, they may have trouble finding their way back to the portal page. The user has to use the back button home button or some other navigation feature of the browser to return to the portal page. While this basic kind of navigation is simple to implement, it prevents the user from being able to work within multiple channels simultaneously. Portals using basic navigation are frequently information portals where a user is specifically interested in researching one particular topic. We did not feel it would provide the most efficient access to services we were planning to deliver via OneStart.

A second approach to portal navigation, referred to as branching, attempts to keep the user anchored to the main portal page. This is accomplished by opening additional new browser windows whenever specific channels or content are selected. This preserves the main portal page in its own browser window. It also allows the user to work simultaneously on separate portal channel offerings. This model is much closer to what we were looking for in terms of allowing for efficient use of multiple Web application services. However, there are significant navigation difficulties with the branching approach. After users select large numbers of channels, they may find a confusing number of separate browser windows open on their desktop. Navigating back and forth among the various windows can become cumbersome because it can be difficult to tell which channel is which when hopping from window to window.

The OneStart team tried to think of other possibilities that would allow for simultaneous use of application channels without causing navigation problems. A sophisticated new W3C-standard Web technique known as I-Frames allows just that. Through the use of I-Frames, portal channels appear as frames within a portal page. The user can work independently and simultaneously within multiple frames on a page using I-Frames. New browser windows are not required and the portal page is not entirely replaced by the selected content. Essentially, an I-Frames portal page is made up of one-to-many Web sessions that are presented as independent components of a single page of portal channels. Each of these channels can be worked on independently of the others without the user ever having to leave the portal framework.

The navigation features are only one of the benefits of using I-Frames. Another benefit is the ability to distribute Web content within the portal. Using I-Frames, portal channels are simply independent Web browser sessions. Therefore, each channel can be hosted, deployed, maintained and supported by a completely separate development team. A simple URL and some basic configuration parameters are all that is required to make Web content a channel in the OneStart portal. This advantage allows distribution of the labor of creating Web content and services. By building the portal framework using I-Frames, the OneStart team was able to utilize all of the Web technology development being done by various groups across our campuses. Given the size and complexity of Indiana University, this allows for a much more rapid growth of the number of services available in

OneStart, which is essential if the enterprise portal is to be adopted widely early in its development.

There are some drawbacks to using I-Frames. The biggest problem is the demanding browser requirements. I-Frames are supported by Internet Explorer 5.0 and above, Netscape 6.1 and above, and Mozilla 0.9.3 and above. While the majority of users of IU Web pages (80-85%) use IE 5.0 or above, a fairly significant number of people prefer Netscape. For users of Unix-based workstations, IE is not even an option. Given that Netscape 6.1 is a very recent version of the browser, a number of IU Netscape users still used version 4.7. Therefore, we had to make a concerted effort to get Netscape users to upgrade to the latest version of that browser.

With or without I-Frames, the introduction of a portal required the evaluation of browser support for enterprise applications. In some instances, different applications were recommending different Web browsers for their particular systems. Obviously, these types of conflicts become problematic for our portal implementation. If the portal supports a specific set of browsers, all applications delivered within the portal framework must support that browser. Therefore, IU was forced to recognize the need to standardize around a specific set of browsers for enterprise applications. Limited resources for support and maintenance of Web applications and the wide variety of browsers and platforms on our campuses necessitated it. We determined to support at least one browser for each major platform (Windows, Mac and Unix). Other browsers may work with the portal but we will take responsibility for resolving only those problems reported by users of supported platforms.

Building OneStart—Initial Development

Once the strategies for the OneStart development methodology, governance structure, user interface and channel navigation were identified, it was time to begin development. Actual coding of the OneStart application began in February 2001. After two releases and approximately nine months of development work, much of the basic portal framework is in place.

A user can log in and be recognized immediately as faculty, staff or student. The campus with which the user is associated is also recognized immediately. Once logged in and recognized by role, the user is presented with a default set of services. The OneStart team continues to work with each role's focus group to determine the default channels for each role.

As of December 2001, portal users must log into individual portal services separately, because of the many different types of authentication currently being used. The OneStart team is working on a mechanism by which users can carry authentication credentials with them wherever they go within the portal framework. This is something our users anxiously await because it will allow them to realize great efficiencies. However, it does open new questions regarding the need for timeouts of portal sessions. Once a user can gain access to all of his or her services by signing on one time, a session in the portal becomes very powerful. Therefore, the university's auditors and policy-making officers may have sensitivities about the timing out of portal sessions. Once it is implemented, the single authentication feature will be one of the most important deliverables to come out of the OneStart project.

Other current features of the OneStart framework include personalization options, which allow users to change the content on their portal pages; modify the themes of the pages; rename, create and delete portal pages; and specify personal preferences. Some portal features include Mobile Bookmarks, My Custom Channels and What's New in OneStart channels. Mobile Bookmarks allow users to import their bookmarks or "favorites" from their Web browsers to their portal pages. Unlike storing these bookmarks on their local PCs, keeping them in their portal profiles allows users to access the bookmarks from any computer with an Internet connection and a Web browser. The What's New in OneStart channel contains a list of the latest portal channel offerings. An online tutorial provides new portal users with a brief introduction to portal terminology and navigation features. My Custom Channels allows portal users to add most Web pages as custom channels on their portal pages. With it, favorite sites that may not be included in the users' default portal channels can be included.

With the framework in place, we are now focused on working with other development teams to develop additional services for the portal. The portal team created a "Channel Developer's Guide." This guide is a very simple set of guidelines for Web developers to follow in order to provide channel content for OneStart. With the framework in place, OneStart provides a service delivery vehicle with built-in customization and personalization features. Eventually, this framework will also include authentication credentials that are carried with the users throughout their portal sessions, providing single sign-on access to the services available. The OneStart framework provides a convenient and integrated solution for access to Web-based services at Indiana University. From the perspective of the service providers, OneStart provides an environment that will attract their specific user audiences, allowing them to find all of the available application services in one place.

As the number of portal services begins to grow, one of the most challenging tasks involves the development of a portal "ontology." If applications are to share information in a meaningful way, a standard set of terms and definitions, or ontology, must be established. What terminology do you use to describe the individual services? How do you organize the list of services in a meaningful way? As the content within the portal grows, this issue becomes more and more important. The ability to organize a very large quantity of portal content to facilitate locating and identifying pertinent information and services is critical. The OneStart project team is working with experts from Indiana University's School of Library and Information Sciences to develop the ontology for OneStart. The basic content of the folder hierarchy that contains the individual Web services needs to be in place early in the process. Once the portal users become familiar with the hierarchy, it will be difficult to make changes without confusing faithful portal users. Likewise, logical and meaningful names for services must be developed so that users may easily find and identify channels of interest.

Building OneStart—Lessons Learned

Perhaps surprisingly, the majority of time spent on an enterprise portal project does not involve technical issues. While there are many technical challenges, the majority of them are organizational. The significant issues raised with regard to policy, ownership, control and priorities can take extremely large amounts of time to resolve. The larger the institution and the more diverse the stakeholders, the more difficult these problems become. It is important not to underestimate the amount of time and resources required to implement the organizational changes required to develop an enterprise portal. These changes ultimately are a way to make your institution perform more efficiently and effectively.

It is absolutely critical that an enterprise portal project get buy-in from the information technology staff. The degree of collaboration and coordination involved in getting an enterprise portal up and running is huge. If the entire IT organization is not working together, a lot of time may be wasted. Buy-in should not stop here. An enterprise portal project must be supported by the institution as a whole. If the goal of the project is to provide a Web portal for all faculty, staff and students, each of these groups should be supportive of the effort. The alternative to getting this support is that other groups may buy or build portals for their own constituents or for their own particular niches. A fundamental rule of enterprise portals is that there should be one and only one portal at an institution. Multiple portal investments would defeat and dilute the benefits of an enterprise portal project. Therefore, the portal vision must be communicated and shared across the enterprise.

A clear and detailed plan of action must be shared with all of the major stakeholders. The best way to get this buy-in is to engage faculty, staff and students in the project early and often. Showing them a working demonstration of the portal concept using "real-life" examples is the best way to communicate the vision. Of course, communication should not stop once the portal project is underway. It should be a continuous process in which ideas and feedback can be shared directly and frequently in order to ensure that the enterprise portal is meeting the needs of faculty, staff and students at the institution.

A formal portal governance structure must be formed to determine the priorities of the portal project. What services must be worked on first in the portal? What requirements are most important for the framework? These kinds of questions need to be answered by the key stakeholders in the project. If the group is very diverse, this can be a difficult process to manage. Groups for each portal role must be formed to identify the default views that each new portal user in a given role will see. What services will be there by default and where should they appear on the page? The institution must get this right the first time because many users will not use the personalization features. Screen real estate is scarce in a portal. Who gets the best location on the first page is a very difficult matter to resolve. The institution will want to consider reserving space for university- and campus-wide news and announcements in order to build a virtual online community. All of these factors must be considered and a formal governance structure will determine the outcome. If the structure is set up properly, all stakeholders should feel a part of the decision-making process.

A number of policy issues are raised by the portal initiative. There are certainly security issues with the implementation of single sign-on for people who have access to numerous applications. How can these services be secured appropriately? Privacy issues are raised and must be addressed. How much data will be kept for each user? How will the data be used? Will the site contain or allow advertising? (The initial response to that question at Indiana was, "Absolutely not!" However, when the project team began talking about an alumni role and potential opportunities for portal users to buy textbooks, theater tickets, athletic tickets and IU T-shirts, the issue was suddenly not so black and white.) Finally, who "owns" the enterprise portal? The scope of the enterprise portal project is so broad, it is difficult to identify one group that is appropriate as the "owner" of the project. This can create communication and control issues within the organization.

The largest technical challenges are the integration issues. Dave Koehler, Director, Office of Information Technology, from Princeton University states this quite succinctly. At a recent conference presentation discussing the future of information technology, he said, "We are no longer developers. We are integrators." This is undeniably true. At IU and elsewhere, institutions are no longer building all of their IT solutions from scratch. For many solutions, perfectly capable third-party software packages are now being sold at reasonable prices. As institutions begin to develop portals with a mix of custom-built software, off-the-shelf vended packages and purchased frameworks customized to meet individual specifications, the integration issues loom large. IU is challenged with implementing this mixed bag of applications in a manner that allows specific information and logic to be shared with other applications. The university is also working to build usable interfaces that insulate users from the complex processing that goes on behind the scenes. These are the solutions that IU's users are demanding and the university must plan for the difficult integration work behind them.

An institution must begin to get its middleware house in order before a portal project can succeed. For example, it must have a global directory for storing information about people. This will be key to all of the institution's integration requirements. The global directory is also a key security strategy. The alternative means having duplicate user data stored in various places across your institution. Many of these servers may be less secure than the institution would like, given that they contain private information about individuals on campus. Having this information centrally located in a secured environment is critical.

IU believes that the integration requirements also necessitate an "open" architecture. Any software, whether developed in-house or by a third party, that requires proprietary platforms or that does not provide standard interfaces into critical business functions, will wreak havoc on enterprise portal development and integration efforts. The majority of vendors today recognize the need for enterprise application integration. Therefore, they are beginning to provide better interfaces into their products. Products are also being developed with a more open mind in terms of the platforms on which they may be deployed. The emergence of the interoperability standards for Web services such as SOAP, XML, UDDI and XSLT hold great promise for integration in the future.

An institution must also deal with accessibility regulations. Frequent usability and accessibility testing is essential. The amount of time required to address usability and accessibility concerns can be significant. Accessibility regulations are simply the law, so campus portals must provide accessible applications for disabled users. Usability is a requirement for all users, especially when it comes to enterprise portals. If a portal is to become the one place to go for access to services, it must be easy to use for people with a broad range of technical background and experience. Only frequent usability testing and iterative releases to correct usability problems can make this a reality. Usability cannot be fixed all at once. It is an ongoing and iterative process that makes the site better with each release.

In fact, the entire campus portal will inevitably be developed in iterations. An enterprise portal is a complex application. The most effective way to make progress is with "baby steps" and multiple iterations. Any attempt to solve all of the problems and issues, while developing a portal with all of the necessary requirements, may be so daunting that it will never get off the ground. Trying to address all of the requirements at once risks "analysis paralysis." Gathering all of the specifications and requirements, including the "pie in the sky" ones, is a good way to start. But those must be prioritized and worked on a few at a time. A good project will build upon its successes. Mistakes are inevitable. Planned iterations give the developers a chance to correct problems while adding new functionality. The enterprise portal project will never be completely finished. It grows and adapts as requirements and technologies at institutions change over time.

FUTURE OF ONESTART

The most important future addition to the OneStart portal framework will be sticky authentication. Indiana currently has no fewer than four separate authentication mechanisms: Kerberos, NT Domain, PIN number, and Safeword Card (token). A fundamental requirement of OneStart is the ability to authenticate to the portal one time and then be authenticated to all of the other services available to portal users. This is the "Holy Grail" of campus portals and it requires the presence of a global directory for all of the individuals at the institution. If the institution lacks such a directory, then it will find itself trying to patch together disparate sources of data about people in order to simulate a global directory. This can get very complicated and may be downright impossible. Currently, Indiana is working on an interim solution until the planned global directory is in place.

Another future addition to the OneStart portal framework is the integrated workflow engine and action list. The ability to automate routing and approval of electronic documents triggered by dates, actions or pre-defined hierarchies is a very powerful feature planned for OneStart. This sophisticated workflow engine is a very important piece of functionality that is missing from most portals. In IU's vision, by way of an action list, each portal user will be notified immediately upon login of tasks that he or she needs to perform. These notifications are not simply e-mail messages. They are electronic documents requiring completion, editing or approval prior to being routed to others or converted into real business transactions. Approvals of Purchase Orders, Personnel Action Forms, and Drop and Add forms are just a few examples of potential electronic documents that could be routed by the workflow engine.

Finally, the OneStart team is also interested in a promising feature known as "adaptive personalization." In the future, the portal will be modified to collect metadata about the behavior of portal users. Depending on what channels a user selects most often and how he or she uses them, the portal could adapt a person's portal profile to these personal preferences and interests. An effective adaptive personalization feature would continue to adapt as a user's interests changed over time. The OneStart team is partnering with Javed Mostafa, a faculty member in the IU School of Library and Information Science who has done research on Web portals and adaptive systems. Dr. Mostafa's SIFTER technologies (Quiroga &

Mostafa, in press) have demonstrated similar kinds of adaptive features with research projects in the fields of medicine and music.

CONCLUSION

An enterprise portal is very costly to plan, develop and deploy. Any enterprise portal project should begin by identifying the problems that need to be solved. In some cases, an enterprise portal may be the solution. It is absolutely critical that a rock-solid business case be made for having only one enterprise portal at an institution. Unified support for the project is a must.

The number of resources and the amount of collaboration required by an enterprise portal project will necessitate many changes within the IT organization. Therefore, deciding to implement a campus portal and developing a detailed plan for getting there must involve all of the key stakeholders. Administrators, faculty, support staff and certainly students are key stakeholders in portal projects. The project will require a continuous investment of time, money and resources in order for the service offerings to grow and adapt to changes in organizational needs and to respond to technological advancements. That kind of commitment requires the broadest support.

For large, complex institutions, an attempt to develop *one* enterprise portal for the entire institution presents tremendous challenges. In order to make the best use of available technology resources, Indiana University's approach for an enterprise portal hinged on the ability to provide a service delivery framework or "vehicle" for the numerous service providers across the institution. With well-defined interoperability standards and open architectures, this distributed model for providing portal content and services can succeed. IU hopes to coordinate Web development efforts on all campuses and within every corner of the institution to implement an enterprise portal approach to service delivery. If we are successful, the faculty, staff and students of Indiana University will have access to the widest possible range of services in a highly usable environment.

By following a few simple guidelines, service providers can quickly develop content for the portal allowing the number of services to grow very quickly. A solid framework must be in place before development teams will be willing to devote resources to making the portal their service delivery mechanism. Once the service providers are comfortable with the portal as a service delivery framework, the enterprise campus portal will provide enhanced opportunities for faculty, staff, students, prospective students and alumni to find the valuable information services they seek. As OneStart becomes ubiquitous at Indiana University, the institution envisions the formation of a virtual online community, creating a global population of life-long IU citizens.

REFERENCES

Frazee, J. P, (2001). Charting a smooth course for portal development. *Educause Quarterly*, 24(3), 42-48.

Looney, M. & Lyman, P. (2000). Portals in higher education: What are they, and what is their potential? *EDUCAUSE Review*, 35(4), 28-36.

McRobbie, M. (1998). *Architecture for the 21st Century: An Information Technology Strategic Plan for Indiana University*. Unpublished report, Indiana University, Bloomington, IN.

Phifer, G, (2000). Enterprise portals—Growing up quickly. Paper presented at the *Gartner Symposium ITExpo 2000, Orlando, FL,* October.

Quiroga, L. & Mostafa, J. (in press). An experiment in building profiles in information filtering: The role of context of user relevance feedback. *Information Processing & Management*.

Chapter IX

Begin with the End (User) in Mind: Planning for the San Diego State University Campus Portal

James P. Frazee, Rebecca Vaughan Frazee and David Sharpe
San Diego State University, USA

ABSTRACT

This chapter presents a case study of the campus portal planning process at San Diego State University. The authors describe the use of participative decision-making strategies that capture the voices of key stakeholders, identify their concerns and priorities, and facilitate a successful portal rollout. Data was collected from faculty, students and campus leaders through a series of focus groups, interviews and online surveys. Findings were examined in light of the literature on technology adoption and the authors' familiarity with portal initiatives at other large public universities. Participants described their vision of the ideal portal solution in terms of features, user interface and functionality. While faculty and students expressed enthusiasm about a campus portal, they also had concerns regarding training, support, reliability, security and standards. The authors make recommendations for addressing user concerns such as providing direction and leadership, segmenting the rollout, communicating the benefits and providing organizational support.

INTRODUCTION

At San Diego State University (SDSU), we are currently midstream in the portal development process; that is, we have begun to experiment with a portal development approach but we have not yet built or implemented a portal. We will most likely choose a solution that combines buying off-the-shelf and building portions ourselves. We strongly believe that in developing a portal solution, decision makers must always begin with the end users in mind. It is their needs and concerns that should drive decisions at every step of the process.

In this chapter, we will:

- Present a rich description of a particular phase in the development of the San Diego State University campus portal, from Fall 2000 to Fall 2001.
- Focus on the planning process and decision-making steps that must go into the portal's deployment *before* roll out.
- Summarize our findings based on data gathered from various constituent groups, focusing first on the needs and concerns of faculty and students.
- Highlight priorities and make recommendations for those whose portal implementations will follow ours.

BACKGROUND: OUR PORTAL VISION

Why a Campus Portal?

Many colleges and universities feel pressured to get a portal up and running (Frazee, JP., 2001). Gilbert (2000) suggests that a college or university must first ask itself why it wants a campus portal. Dynamic and individualized Web systems are essential for institutions of higher education and as customer expectations grow, these institutions must further develop their Web-based technologies to distinguish themselves from their competition (Connolly, 2000). Today, faculty, students, staff and administrators have access to an ever-increasing number of databases and ways to interface with them. Most people have multiple usernames, passwords, login locations and interfaces. Furthermore, it is difficult to disseminate information to targeted groups, and duplication of effort and inefficiencies plague the storing and retrieving of some information. The portal aims to simplify these processes, improving the efficiency and effectiveness of a number of functions by making information available through a single access point. The overarching goal is to improve communication and engender an increased feeling of community.

Our Definition and Vision of a Portal

At SDSU, we define a portal as a customizable entry point into campus-wide administrative functions, information, resources and Web-supported courses, using a single username and password. Essentially, we think of a portal as a tool that allows people to organize and customize their working environments. Users will be able to control their views of the Web, designing the look and feel of their portal pages in ways that make sense to them, controlling the information they receive and how it is displayed using a publish-and-subscribe model. For instance, a student could easily publish a change of address or a change of major or subscribe to a service that provides notices of campus concerts and events. A portal should also include an online calendar that users can modify to allow events to be automatically entered and updated by the system; such events might include course meeting times, deadlines, exams, and links to course assignments and readings.

Our vision is that the portal will be the first place that the SDSU community member goes when turning on the computer. Ultimately, the portal will not be seen merely as a tool. Rather, it will be a resource that is transparent and seamlessly integrated into daily activities, enabling the end user to achieve their academic, research and community goals more effectively and efficiently.

Why Begin with the End (User) in Mind?

"Begin with the end in mind" is one of the "Seven Habits of Highly Effective People," according to Dr. Steven Covey. In his seminal book by the same name, Covey describes many powerful lessons for leading change. Drawing from the words of business gurus such as Drucker and Bennis, Covey suggests that "management is doing things right; leadership is doing the right things" (1989, p.101). It is easy to grasp the significance of this idea when we consider the portal planning efforts at SDSU. When choosing a portal, as managers we may be overwhelmed by decisions about hardware, software, staffing and related costs. However, in order for any campus-wide solution such as a portal to be successful, we must step back and act more as leaders, first making sure that we are doing the right things. This is where analysis comes in. Analysis involves partnering with those affected most by the change, gathering several perspectives and proposing solutions based on data, not habit (Rossett, 1999).

The purpose of this chapter is to describe how we embarked on the portal planning process, carefully considering where we were and where we wanted to be *before* forming opinions on how to get there. In this chapter, we will show how we went about gathering the shared hopes and fears of the SDSU academic community, and how we analyzed and aggregated their opinions into themes that will shape our thinking as we move forward with the development of our portal.

PHASE ONE: PORTAL DECISION

Focus on Academics

We knew we needed a portal, but we didn't know exactly what would give us the biggest "bang for our buck." In the summer of 2000, SDSU began the process of determining the focus for our campus portal system. University leadership formed a 16-person *ad hoc* committee to evaluate and make recommendations regarding various software vendor solutions. The committee included members from Academic Affairs, Alumni Affairs, Associated Students, Athletics, Aztec Shops (campus bookstore and dining services), Business Affairs, Student Affairs, University Advancement and the University Foundation.

We decided that the SDSU portal would focus primarily on supporting the academic mission of San Diego State University by enhancing a sense of community and facilitating communication among the SDSU constituencies. The portal would provide one point of access to information and resources that are key for teaching and learning as well as those that are essential to administrative operations. The portal would function through an easy-to-use Web-based environment that is user customizable and ensures security of all significant data.

Guiding Our Efforts

During our initial efforts to determine strategies for portal implementation, we turned to the change management and technology adoption literature (Ely, 1999; Frazee, R.V., 2001; Fullan, 1999; Hall, 1987; Hall & Hord, 1987). We identified several success factors that would serve as benchmarks for our own efforts, and focused on the following at this stage:

- **Identify Key Stakeholders**: Involve all those *affected by* the portal and those who *can influence* the outcomes of the portal, including prospective students, students, faculty, staff and the greater SDSU community (e.g., alumni, donors).
- **Identify Stakeholder Concerns**: Become familiar with their worlds (priorities, goals, historical barriers, political realities) and anticipate questions and skepticism. Ask prospective students, current students, faculty, staff, alumni and community members about their issues (e.g., time, access, security, privacy).
- **Involve Stakeholders**: Share decision making through stakeholder steering committees (e.g., the *ad hoc* portal committee). Collect input through interviews, focus groups and surveys. Target opinion leaders and resistors, using multiple communication channels.

Shared Decision Making

To plan strategically, we must not only collect perspectives from various parties throughout the organization, we must also lead a decision-making process that facilitates communication and participation and helps all parties feel bought-in (Ely, 1999). There are many campus constituencies, including students, faculty, and staff, who can contribute to and benefit from successful implementation of the portal. At the same time, these groups will also need to make changes in their standard operating procedures to accommodate and fit into the new portal system. For example, while a portal will provide benefits for the SDSU community, it will also ask a decentralized organization to standardize some operations. Therefore, in order to take full advantage of the portal's new options for communication and information distribution, we must engage each of the constituent groups as early as possible in a participative process of decision making and implementation (Gilbert, 2000).

Developing a Rubric for Software Selection

The *ad hoc* portal committee developed a list of desired features, identified potential portal products, reviewed information concerning each product and selected those that met minimum criteria. To guide our efforts, we started with a list of features that evolved into a quantifiable metric, or rubric, for evaluating portals (Frazee, J.P., 2001). The portal rubric includes criteria for scoring several aspects of the portal such as usability, features, IT/management issues and business issues (see Appendix A). Our commitment to a collaborative process ensured that every key group had a hand in the creation of the rubric.

So, What Did We Choose?

In Fall 2000, we reviewed a half dozen vendors using the rubric. We found that vendors make numerous claims and that few campus-wide portals are actually in full deployment. There are two main types of portals: business and instructional. The committee decided that neither had all the capabilities that SDSU desired. As the Director of Instructional Technology Services (ITS) explained, "something you can buy off the shelf isn't as far along as it needs to be, to be easy to use and have all the features we need."

Therefore, we are experimenting with a solution that would take advantage of existing software already in place at SDSU, specifically Oracle and Blackboard. We expect to use a hybrid approach that will attempt to integrate the Oracle student information, human resources and financial systems and the Blackboard Web-based course management system. The user interface will be built using Oracle development tools.

To begin with, the portal's focus will be academic. For instance, priority will be on helping faculty and students easily communicate with each other, and linking class lists and grades between our Oracle student information system (SIMS/R) and Blackboard courses. We anticipate that emphasis will also be on providing instructionally related links and resources, as well as supporting research committees and other groups in their collaborative work.

PHASE TWO: GATHERING USERS' NEEDS AND CONCERNS

> It's 11:58 AM and the pizza has just arrived in the ITS conference room. The research team continues readying the room for a group of students from Associated Students. Six wireless laptops are booted up with browsers pointing to the online portal survey. The data projector is fired up with the PowerPoint presentation that will be used to present a vision of the SDSU portal. The video camera is ready to record the whole session. The goal is to generate discussion about what students want and don't want in a campus portal, as well as to gather feedback on the beta version of the survey.
>
> A member of the research team asks these students the same questions as were asked of the first focus group earlier today. "Think about what we've just talked about. What do you think we should focus on first?"...
>
> As the focus group is wrapping up, the research team asks for any additional comments about the portal. "It sounds great." "I can't wait." "This will be great...if it works."
>
> Meanwhile, another member of the research team concludes her telephone interview with a member of the Nursing School faculty. "What do you see as the biggest opportunities for the portal to help you in your work?" "Making accessibility to research easier. If I knew that I had something that was dependable, and that's a big word here, I would certainly want to use it."

Our Methods

We decided to take a phased-in approach to developing and implementing a portal rather than attempt to build a complete portal that would meet all of our goals at once. We decided to focus on faculty members and students first; later we will examine other constituent groups including staff, alumni, prospective students and the San Diego State community at large. Our next task was to answer one overarching question, "Where do we begin?" What do faculty members and students want? What concerns them the most? How do we build awareness of the portal among these groups? How do we concentrate on the primary needs of these end users to ensure that the portal rollout is a success? We targeted campus opinion leaders including officers from Associated Students (our student governance organization) and members of the Faculty Senate Instructional Technology Committee. They were charged with helping us identify users' main wishes and concerns for the portal. We conducted interviews, focus groups and site visits. We developed an online survey that we pilot tested with faculty and students. We plan to distribute the survey via e-mail to all faculty and students.

Our Research Team and Methodological Approach

We began phase two of the planning process in Spring 2001 with interviews, site visits, and a literature review. In September 2001, we developed a research team including the two co-authors of this chapter, James Frazee, Associate Director of ITS, and Rebecca Vaughan Frazee, a doctoral candidate in educational technology, plus two master's students in educational technology.[1] We drafted a research plan for using multiple data collection strategies to obtain information from multiple sources. We limited our study to faculty, students and campus leaders. We also bounded the study by time (12 months), consistent with a qualitative exploratory case study design (Asmussen & Creswell, 1995; Yin, 1989). We selected this approach because existing theories and models were not available for assessing campus reaction to a portal.

Following the constructivist tradition, we chose to analyze and illustrate our findings in the context in which they were uncovered. To name the categories in the narrative, we employed an in vivo codes approach, which uses direct quotes to capture the voices of informants (Strauss & Corbin, 1994).

Bracketing Our Perspective

Our interpretation and presentation of the data must be considered in light of our own perspectives and methods. First, the survey was skewed to include disproportionately more graduate students than undergraduates. Secondly, because of the personal nature of focus groups and the rich discussion during interviews, those opinions may seem more emphatic than opinions voiced through the anonymous survey. Finally, as representatives of Academic Affairs, our research team is admittedly focused on academic issues.

> As Steven Covey, the inspiration for the title for this chapter, writes: "The more aware we are of our basic paradigms, maps or assumptions, and the extent to which we have been influenced by our experience, the more we can take responsibility for those paradigms, examine them, test them against reality, listen to others and be open to their perceptions, thereby getting a larger picture and far more objective view." (1987, p. 29)

As researchers we must bracket our biases. We have attempted to set aside our own "prejudgments." We used what Creswell (1998) calls "member checking," by taking a preliminary draft of our findings back to the people with whom we spoke (e.g., the leaders and faculty we interviewed) in order to have them verify the accuracy. As you will see in our findings, we used triangulation to converge the multiple sources of qualitative information and quantitative data in order to develop themes emerging from participants' hopes and concerns. Finally, we provide interpretation or "assertions" of the lesson learned, and couch them in terms of the literature (Lincoln & Guba, 1985; Stake, 1995).

Participants

We included a total of thirteen SDSU students and four faculty members in focus groups during November 2001. We interviewed two faculty members via telephone and conducted two face-to-face interviews with campus leaders who are directly involved with the portal: the Director of ITS and the Director of SIMS/R (Student Information Management System/Relational). We also collected online survey responses from thirty-five anonymous students, both undergraduate and graduate, from four of the seven colleges that comprise the university.

We used purposeful sampling to select focus group participants so that we would be sure to include opinion leaders from faculty (Faculty Senate Instructional Technology Committee) and students (Associated Students). In addition, the SDSU administrators were purposely selected because of their roles on the campus and how they relate to the portal. Convenience sampling was used to enlist eight readily accessible student workers from the ITS department and two faculty members who had had recent interactions with ITS.

Instruments & Procedures

Online Surveys

We developed an alpha version of an online survey based on the literature and our experiences developing the rubric mentioned earlier. Before piloting the survey, the Associate Director attended the 2001 EDUCAUSE Conference,[2] where he spoke with Dr. Carl Berger and learned of similar efforts at the University of Michigan (UMich) where they had developed online surveys as part of a broad-based strategy to collect input from students and faculty (personal communication, 2001). Dr. Berger encouraged us to freely adapt the surveys for SDSU. We developed a beta version of our survey that included a small section for demographics and computer use, and focused mainly on questions about concerns and preferences for using the Web for academic, administrative, communication, personal and miscellaneous purposes. The survey started out with more than one hundred items and was whittled down to forty-eight, based on focus group participant feedback. We plan to distribute the final, much condensed version of the online survey to all SDSU faculty and students (see Appendix B).

Focus Groups

We conducted four student focus groups and one faculty focus group. Each focus group lasted approximately sixty minutes. Each session began with a brief presentation about the portal including our vision, process, what we've done so far. We then had participants pilot test the online survey. As students and faculty completed the online survey, they were asked to think aloud. Their comments about the survey prompted conversation. Finally, we asked several open-ended questions about their wishes and concerns regarding a campus portal, and what the university should consider when rolling it out to faculty and students. The general questions we asked included:
- How would you describe the perfect campus portal solution?
- What about the portal is most promising?
- What concerns you the most when you think about the portal?

Clarifying questions were informal and built upon the conversations that emerged.

All focus groups were videotaped. Faculty completed the online survey in their regular meeting room using computers located there. They were also given hard copies of the survey to record anonymous comments. Based on feedback from the faculty, we further refined and shortened the survey. In subsequent student focus groups, participants completed a revised survey using computers in the ITS conference room.

Interviews

We interviewed two faculty members by telephone. The interviews focused on their wishes and concerns about a campus portal. Each lasted approximately 60 minutes. One faculty member was from the College of Health & Human Services's School of Nursing, and the other was from the College of Education's School of Teacher Education. A semi-structured interview protocol was employed that asked seven questions, including:
- How might the portal help you and other faculty members the most?
- What must SDSU do to successfully launch a portal?
- What should we be concerned about when rolling out the portal to faculty?

Face to face interviews were conducted with two university leaders, the Directors of ITS and SIMS/R, in their offices. Each lasted approximately sixty minutes, and the focus was on allowing them to voice their perspectives relating to questions such as:

- How does your position relate to the proposed campus-wide portal? What are your main responsibilities and duties?
- What concerns you personally as we plan what must be done to implement the portal?
- What about this change is most promising (benefits)?

Learning from Other Institutions

We also wanted to know first hand how other campuses were handling decisions about their portals. Did they build, buy or choose a hybrid? How were they rolling it out? Who was involved in the decision-making process? What lessons could they share with us? To find out, we talked to decision makers from several universities in various stages of portal implementation. We participated in portal listservs (e.g., the EDUCAUSE portal constituent group). We visited the University of Tennessee to meet with their Assistant Vice President of Educational Technology and the Manager of their course management system. We conducted several conference calls with current portal administrators around the country including those at Arizona State University and the University of Oregon. We attended portal conference sessions and special interest group meetings at local and national conferences (e.g., EDUCAUSE 2001, CONVERGE 2001, SYLLABUS 2000), gaining valuable insights from institutions such as the University of Michigan, NYU, Cal Poly San Luis Obispo and others. We visited portal websites that are up and running (e.g., UCLA, University of Washington, University of Texas).

FINDINGS

> Myra: I think this is a good thing, because I HATE having so many passwords.
> Gabby: I have at least five passwords, just on campus.
> Cliff: Yeah, I like the idea of having one password for everything, because I've got one for financial aid, one for career services, Blackboard, e-mail...And I've already forgotten the passwords, so I just don't use the services anymore.

Benefits

Convenience

All participants agreed that the portal could simplify the lives of users by improving communications and saving time. One leader stated that "SDSU strives to be a cutting edge institution, and there is a growing expectation on the part of students who like to apply electronically and track the process without having to call or come to campus just to wait in line." Using the five categories from the Stakeholders Benefits Matrix (Table 1),[3] we found that for participants, the biggest benefit of the portal seems to be convenience.

Table 1. Portal Benefits Matrix

SDSU PORTAL: BENEFITS FOR STAKEHOLDERS					
Benefit	**Prospective Students**	**Students**	**Faculty**	**Staff**	**Alumni / Community**
Convenience	Peek into the SDSU community by discovering courses prior to enrolling, as well as, applying and checking status of application online	Access course materials, academic calendar, final exam and test dates; request an unofficial transcript and apply for course forgiveness	Create courses that infuse campus and Web-based resources with automatic student enrollment in course management system (Bb) based on information provided from student info system (SIMS/R)	Review and update information vital to campus operation including student enrollment numbers, revenues, expenditures and campus map	Monitor university athletic, music and drama events anytime, anywhere, as well as what's new on campus, based on personal interests
Cooperation	Share enrollment experiences with staff and other prospective students	Collaborate with other students through online communities	Build relationships with peers intra- and cross-departmentally to create resources for students	Work with students, faculty and peers to support the teaching and learning process	Keep in contact with fellow alum and get involved in alumni activities
Communication	Live interaction through real-time chats and discussion groups	Stay connected with faculty and fellow classmates	Keep in touch with students about enrollment, assignments and grading	Stay abreast of changes in campus events, schedules, policies, etc.	Stay alert to class notes and career resources
Capacity	Streamline application process through improved service	Post assignments and review grades; change major or home address	Increase productivity by iteratively improving and recycling course materials	Update information quickly and easily without going from building to building	Learn about career fairs, counseling, workshops and job listings

"As far as [the portal is concerned], what I really like about it is, I would have access to everything I have to do on campus…and I would have access to it at my finger tips. And what I would really like to see would be 2-4 hour access to it."

Communication

Communication was another benefit cited by students, faculty and leaders. Leaders felt that the portal could improve communications, which could save time and, ultimately, improve instruction:

"It can increase communication, and simplify some tasks, hopefully a lot of tasks. Build community with students, alumni and prospective students. Possibly because of simplifying peoples lives, and making it easier for faculty to communicate with their students…, it could possibly improve instruction, although I think that more of a spin-off benefit…. [The portal can provide] help for students, communications with different student organizations. [There is] potentially more buy-in from the different student groups because of better communication."

"Institutional Information and Services That Simply Aren't Available at All Elsewhere."

What would make the portal "stick?" That's what was asked of participants of the EDUCAUSE portal listserv. The Director of Research and Development at Wake Forest University responded this way:

> "Our campus constituents visit it because it's the most convenient way to look up grades, register for classes, find someone, submit payroll timesheets, register a vehicle, purchase textbooks from our bookstore, e-mail an entire class, etc. not because it's *sticky*. We aren't entertaining or advertising, we're providing institutional information and services that simply aren't available at all elsewhere or are much faster and easier through our portal than through other means."—Anne Bishop, September 5, 2001

We also asked our focus group participants what would keep them coming back to the portal? Surprisingly, students echoed Bishop's perspective. Josh told us, "As a student, I can get news and I can get sports and I can get all that stuff other places. The most important stuff first would be the more SDSU-based stuff." However, Eric, a computer science graduate student, felt that in order for students to make the SDSU portal their main default Web page, it would have to offer more. "If it's just all academic, they're just going to turn in their homework. But if they can personalize it with all that other stuff, (local, national, international news; entertainment etc.) they're going to keep going to it."

Quality

According to the people we met at UT, their portal has provided "real streamlined access to information." They say the benefit is quality, not quantity. Rhonda Spearman, the course management system manager at UT, thoughtfully shared a matrix of benefits and other project management resources that stressed the need to identify and inform stakeholders. Their approach has been one of inclusion, and the process seems to be paying off.

What Do They Want in the Portal?

During the student focus groups, we showed participants our idea about what the portal might look like, and we toured them through a few existing university portals. We asked these students, as well as faculty and leaders, to imagine the "perfect" SDSU portal. What would it include? What might it help them do that they currently could not accomplish, at least not without a hassle? What were "must haves?" What should we make sure to avoid? Their discussion centered on features, user interface and functionality. To find common themes, we compared their comments with items and categories that emerged as priorities from the survey.

Ease of Use

Keep Websites Consistent Throughout the Campus. As expected, faculty and students said that the portal must be easy to use and have a "seamless interface." We asked participants to think about what they liked and didn't like about existing campus websites. In addition to common principles of good Web design (e.g., enough blank space, intuitive icons and navigation), their comments seemed to revolve around one main theme: consistency. One survey respondent said, "make all course websites conform to one template to make them easier to navigate. Also, courses should not use software that does not cross platforms!" Dr. M. echoed the need for consistency as follows: "I think it is just amazing at this university that we don't have a standard template across the entire university to be doing all these things…We don't even have programs that communicate with one another across campus, from department to department, or college to college…If this whole concept [of the portal] could promote the idea of more uniformity and some templates that were flexible, something that you could grow very familiar with, to me it makes perfect sense."

Make It Dummy Proof. Not surprisingly, one of the main selling points of the portal is its customizability. However, while most students said they like the idea of customizing their own portal homepage, it's questionable just how much effort users would be willing to expend modifying a portal homepage. Some students voiced concern that there might be some users who "don't want to play with it." In fact, when we asked about the commercial information portals that students are currently using (e.g., My Yahoo, My MSN), even Chris, a computer science graduate student and campus tech coach, admitted, "I don't like it but I haven't really taken the time to change it." Nick went on to add, "Some people are very computer illiterate, so if they go to set up one of these accounts, I think you should have a real dummy-proof way of setting it up…The first few times, if they can't use it, they'll just bag it."

Perhaps, as Eric suggested, the portal should offer an interface with "different levels for advanced people to customize," and a simple default page that doesn't require any customization or set up "for people who don't want to do anything."

Administrative Functions

Not surprisingly, students and faculty were more interested in using the portal for administrative and academic purposes rather than for personal interests. When surveyed, students strongly agreed that the top five administrative uses for the Web were to get information on courses prior to registration (100%), find out days and hours of operation for all campus offices and services (97%), register for courses (94%), check progress toward their degree (94%) and order transcripts (91%). As one student put it, "anything that has to go through the cashiers." Clearly, these areas warrant attention from those designing their own campus portals.

"My Web Advisor"

Participants definitely want access to personal, academically related information online such as grades and transcripts. Students in one focus group engaged in a lively discussion when someone introduced the idea of a decision support tool that could help them with degree-related issues.

Nick said, "It would be cool if you could type in 'this is what I want to major in,' and it takes you to that page in the catalog and shows you what classes you have to take." "This one's going to be really good," added Annie. Faculty members agreed. "That all is a wonderful thing. As a graduate coordinator and advisor, I long for the day when students have access, and from what they tell me, they do too."

Personal Employment Information

Staff and faculty might want personal employment information. "As a state of California employee, [I would like] access to my info having to do with my personal employment, retirement, etc. at SDSU."

Tools

Students and faculty members want access to tools for communication, instruction and personal productivity.

Access to Site Licensed Software. All survey respondents agreed that they would like to use the Web for conducting research. This included the ability to access a personal file storage area for storing files and "projects." Furthermore, a faculty member and student said they would also like access to software applications via the Web, such as C++ and "data analysis" tools, so they could work on their projects "anytime, anywhere." Over half (66%) of those students surveyed indicated that they would like to use streaming video and/or other advanced technology applications. One faculty member, Dr. M., also mentioned the desire to access instructional development tools. "If I knew there was one place I could go and easily find instructional development tools for Web-based instruction, and different communication options that I could easily set up, all that kind of stuff, I think that would be extremely useful to me."

Targeted Information "Push." Over three-quarters of students surveyed reported that they would like to use the Web to automatically receive announcements from their school or college (88%), or receive automated reminders for assignments, appointments and events (91%). When we described for focus group participants how the portal might be used to send out (or "push") targeted information, students and faculty members were very excited about the possibilities. For example, one faculty member described the difficult task of making various student groups aware of scholarship opportunities, and that "being able to target certain subsets would be great." One student thought that the portal would be much more effective than e-mail for distributing announcements about campus events. "You can get e-mail lists, but folks might not check their e-mail all the time. I can see how a student might check [the portal] every morning, and even several times throughout the day. Check it and see what's going on. It could be a real good source of information."

"The First Thing I Do Is Check My Mail." Students want the ability to access their e-mail via the Web. Furthermore, some students said they want to be able to organize and access various e-mail accounts through the portal, without integrating them. As Cliff said, "I don't like to mix my school and my personal accounts." Several participants and 83% of those surveyed would also like to use the Web to access their personal calendar, course schedule, to-do lists or address book from any computer. One faculty member added that she would want to be able to separate out specific calendar entries, such as those related to her department versus the college or campus.

"News That Might Affect Students." The majority of survey respondents did want information on events (83%), and several focus group participants said they wanted information about campus events, such as concerts and the football schedule, as well as links to the campus newspaper. However, participants emphasized that they wouldn't use the portal as their default homepage if it only contained campus-related information. For instance, students want information on local, state and national issues "that will directly affect the students here." Also, they would like to have links to things like weather, sports, stock quotes and even games.

Interestingly, almost half of the students (46%) reported that they would not want to automatically receive news from the SDSU newspaper (i.e., "pushed"). This could be an indication of how opinions vary regarding the desire to "push" or "pull" various types of information.

Main Concerns

Everyone with whom we spoke noted the value of the portal and expressed enthusiasm about its introduction. Participants also voiced several concerns regarding the portal. This does not come as a surprise since our questions were focused on uncovering their needs and concerns.

Training, Support and Access

"It's Really a Resource Issue." There weren't many surprises here. All those who participated in our research expressed concern about having the necessary resources. One leader felt that the resource issue was a big part of why the campus wasn't farther along with the portal. "There is so much, you have the hardware, the programming, you've got the going out and surveying people to see what needs to be done [first], you've got the evaluation, you've got the training of people to use it, you've got the help desk for when people are having problems. Probably supporting the higher administration when they want to put an announcement on it, you've got someone who is deciding who's authorized to do what… Keeping things up-to-date, who is going to do that?"

"How Will I Be Trained?" Several students asked how people, especially for those who are less computer literate, would learn how to "set up" and use the portal. One faculty member suggested that portal training be mandatory for staff and faculty. She said that training "actually needs to be a part of the education of faculty and staff that is mandatory…I think it has to be a priority so that people know how to use these things so that we can drop some old ways."

At the University of Tennessee, one of their main goals regarding the portal is to provide course management system (CMS) certification for every instructor. These certification courses are all available face-to-face as well as "anytime, anywhere" through the Web-based interface. As the VP of Education Technology at Tennessee said, "practice what you preach." One strategy Tennessee uses for professional development is what they call "lucky sevens." Whenever seven or more faculty members collectively request one of the training sessions from the course catalog, the course is provided "on demand." According to Spearman, this customer-oriented approach seems to be helping faculty who "really struggle with the concept of dealing with multiple windows, and cross-platform issues."

"Will Tech Support Be Available?" While the availability of training was a concern, some students felt that many users might not have the time to attend a training session and would instead want some sort of technical assistance, either online or over the phone. Furthermore, this support would need to be on-demand, with little wait time. Users don't want to wait a long time to receive an e-mail from an online "help desk," and they want to speak to a real person when they call. Myra said, "I know that the most frustrating thing is that when [students] call for any kind of help on campus, there's always a voice mail, there's not a real person."

"How Would It Be Funded?" Students and faculty appreciated the fact that a large, campus-wide effort such as the portal would require money, and they all were concerned about where that money would come from. Ronnie expressed a common student concern, "We wouldn't have some sort of technology fee or something? Because that's what I'm worried about."

While they don't want to be charged additional fees for using the portal, students and faculty don't see advertising as a funding option, and they supported our decision not to allow advertising on the portal. Dr. G. admitted that targeted advertising or customer "profiling" on some commercial portals is especially annoying. "I'm going there because I want a particular thing and they try to push things on me. It irritates me. I don't want [them] to push things at me."

"Faster Access, Larger Bandwidth, More Computers." Not surprisingly, students were concerned with how easily they would be able to access the portal. They talked about having more computers available on campus and in the dorms for access to the Internet. A few survey respondents expressed the need for "easier connections on campus for people with laptops—wireless or otherwise." They also reminded us that many students are still using dial-up modems from home, so pages should not take too long to download.

"One-click Guest Entrance Without a Password." The concern about access also extended to include the discussion of *who* exactly would have access. With a portal in place at SDSU, several faculty members wanted to make sure that the university maintains some level of unrestricted access to the main campus Web pages for the public community at large, access that would not require visitors to establish a password or give out any personal information. Dr. H. asked, "Is there going to be a place where you can just come and look around and say, 'Oh, okay, this is SDSU,' without having to go through a password system?" Dr. J. added, "How much information are they going to have to give to log into the portal? Because that's a stop to a lot of people. I'm thinking more in terms of just community image."

Dr. M. felt that the portal could be a great public relations tool for the university: "It would allow [community members] to enter into a dialogue that they hadn't before." Dr. G. took the idea of community access one step further, feeling that community access should be one of our top concerns, especially access to library resources.

"It should be remembered that we are here to support the community. The university exists for the betterment of this community, therefore it should be available to everyone in the community. Everyone should have access including to the library. Of course that interaction fuels itself too. You develop this synergistic teaching, research, and community service. It could be a tool that would just really magnify leaps forward."

"Will It Interface with Blackboard?" Students and faculty both inquired how a new portal would integrate with existing campus systems. Students wanted the portal to interface with SDSU's existing course management system (i.e., Blackboard) and other Web-based campus information systems. Annie wondered, "What's going to happen with Webline and all these other websites for Financial Aid or Admissions and Records. Are they going to collaborate? They're all going to work together after you do this?"

Dr. H. asked, "Let's say a college or a department wants [its own] portal. Would there be options for that in this portal?" This question is one with which we at ITS are highly concerned. The longer it takes for a central portal to get off the ground, the more likely it is that departments and colleges will install their own versions of "portals" that will then need to be considered and integrated into any future campus-wide portal effort.

Students were also concerned about how the portal would integrate with their personal technology-based systems such as their e-mail and PDA applications. When asked about her greatest concern regarding the portal, Myra said: "Now, on some of this stuff, I use MeetingMaker, I use Eudora, because I'm a staff person too. Would it integrate with those? I have several passwords for different programs, would I be able to access those with one password through the portal or no?"

We would like to add one caveat here about passwords. In our discussions with potential users, we have been including as one of the benefits of a portal the fact that users will only need one password to access many different systems collected in the portal. However, we realized that this may be confusing to some people. We must make sure NOT to imply that ALL possible passwords will be integrated. Only those having to do with campus resources can be integrated into one through the portal. In our case, that includes systems such as SIMS/R, Blackboard and campus e-mail accounts.

Standards and Expectations

"Are You Going to Make Faculty Use It?" Faculty and students want to know that all students have the same chance to conveniently access the same information that other students have when their professors are using the Web. One student expressed his frustration with the inconsistencies in the ways that faculty are using the technologies currently available. "I have this professor that is using something from 'Courses at Yahoo.com,' and its very annoying. It does the basic things that a service [like this] should do, but it's not as good as Blackboard. Like, it doesn't have the discussion part."

Annie admitted that she would be frustrated "if another student gets access to all her homework and syllabus [online] and I don't get that because [my professors] think they're too smart and don't need it." Likewise, Dr. M. wants to know that her students are getting equal opportunities.

"I should be able to say to my students, regardless of what class they're in, regardless of who their instructor is, I should know that they are getting certain pieces of information, or being able to access things on the Web...It should just be that natural. I say these things, and I realize that's a paradigm shift for a lot of people and certainly for the university in total."

"Can We Expect That Everyone Will Be On?" Interestingly, faculty members and leaders compared the introduction of the portal to their experiences with campus e-mail. One leader felt that in rolling out the portal, users must be given options and time to adjust. "Changing from regular mail to e-mail is a gradual process. We'll need to make it optional for students to choose from e-mail, regular mail or pick up." A faculty member agreed: "I kind of wonder about some of the communication content here. Just as an example, when a dean sends out this announcement that says, 'hey we've hired this new faculty member...,' well I'm sure for a long period of time it's also going to come out via e-mail. So, we're going to have this sort of a duplication of effort for some time."

However, Dr. M. urged that in order for the portal to succeed, it must be seen as an expectation, not merely another option. "I think that has to be an expectation, it's not just 'use it if you want.' I think in order for this to succeed, people have to buy in and say, 'Okay this is not going to just be some nice thing that's attractive and useful for some faculty who happen to like technology. This is something...that is going to replace something...' I think if we're going to move in that direction, we have to realize that instead of giving faculty the option to either check their e-mail or check their mailbox, we have to say 'this information is important and you're going to get it via e-mail. We're supplying you with all the equipment, we're supporting you with the training, but that this is also an expectation."

"Is It Condoned by the University?" Faculty members, students and leaders all appreciate the benefits of the portal in terms of efficiency and convenience. They want "enough flexibility so you can use it as you want to." But how much flexibility should be allowed? What about content that is potentially offensive, distracting, annoying or in direct competition with other university entities such as the campus bookstore? When we presented participants with our concept of the portal, including the idea that the portal could contain channels for personal interests such as the weather, stock quotes and selling things, at least one member from each group expressed concern about how we would "control" content on the portal. What would and would not be condoned, and who would make those decisions?

Eric said, "when you customize it, you're probably restricted in some way…you can't just put everything on there." Annie asked, "are you going to have Internet police?"

Interesting issues were raised regarding control and censorship in areas such as pornographic content and the ability to use the portal to sell personal items (e.g., used textbooks, furniture) or promote a "non-educational business." Annie admitted, "if I saw that [I could advertise my business on the portal] — I'm a salsa instructor, so I personally would want to promote myself on here."

One leader brought up an important question about different restrictions for different portal users. "If you make it really easy for some staff to look at their stock quotes, because it's always right on their portal site, then maybe they'll spend more time looking at that instead of looking at what they should be looking at. Although, already it is fairly simple to set this up on your own, but when it comes up on their official Web page, then is it condoned by the university? That's an interesting thing…I don't think I've ever thought of that before. Do you control those kinds of things? Do you let the President check his stock quotes and not the secretary or student? With the portal, you can choose which things (channels) that you want to make available based on their role. That would be an interesting political thing."

"Do Standards and Centralization Match Current Campus Culture?" SDSU leaders describe our campus as having a very decentralized culture. The leaders we interviewed felt that the autonomy of individual departments is well respected, and they doubted the portal would change SDSU's decentralized culture.

"The campus is decentralized. That's going to be one of the biggest and hardest things for pulling this together, is getting people to agree on how to do it, and to give up their control of their data and decide on what would be the best way for people to access data. What levels should people have access to without having to do multiple sign-on with multiple passwords? Does it fit the climate? Not really. Can it be done? With a lot of work, a lot of talking and negotiating, and probably some edicts as to how it will be done."

However, one faculty member felt that some form of centralization or standardization is needed regarding tools and resources, "particularly for administrative kinds of things," and sees the portal as a step in the right direction.

"I think it's going to cause a ripple effect, even without the whole portal idea. The idea of using some uniform tools, administratively and so forth, across all colleges and all departments is going to be a challenging endeavor for the university. But if the university really wants to be able to afford this and maximize their opportunities, they've got to do that."

Personal Concerns About the RollOut

Change, even for the better, can be messy and is almost always a bit uncomfortable. We considered Hall's (1987) Concerns-Based Adoption Model (CBAM), particularly the Stages of Concern.[4]

Awareness, Information and Satisfaction

"How Are You Going to Pass the Word to Students About This?" Students and faculty were concerned with finding out about the portal and how they would get training and support. Annie asked, "How will I learn that this is available to me? What if I don't go to an orientation, who will tell me?"

"Will It be More Work?" Chris, a staff member and graduate student, likes the freedom that students will have with the portal. "The beauty of this is that students could look at their own information, review their transcripts and make sure that their grades got posted instead of having to wait." Participants were also aware that with this freedom comes additional responsibility, perhaps even changes in roles and what is expected of both students and faculty. Dr. G. admitted, "Once you have access to all that information, it's enormously liberating and an enormous burden. My role would be much more global in perspective and much more would be expected of me because I have so many more resources." Students were sensitive to the fact that some faculty might see this technology as simply more work. One student said, "Faculty get scared, they don't want to deal with technology, or maybe it's an increased workload, it takes more time to add stuff and put stuff online…they might say 'yeah, this is great, but I don't have time.'"

John, an undergraduate finance major, thought that faculty should welcome the portal because it places more responsibility on the students themselves.

"I think once they see that it gets a lot of trouble off of their heads, I think they will actually like it. Because it's very annoying when a student comes along and says, 'Hey, I lost my syllabus, can you print me another one for me,' or 'hey, I couldn't do my assignment because I wasn't there and I didn't know.' But as long as it's posted on Blackboard, [faculty] could say, 'Hey, I posted it there on [the Web], so it's your responsibility to do it.'"

"It's Really Impersonal." Though only mentioned by a few participants, one important concern was that of whether the introduction of a portal would somehow detract from the human side of the college experience. "Some students," said Eric, "don't even want to touch a computer." Jason added, "I wouldn't like to see this go too far to where every instructor is using this to get info to their students, or to have homework assignments up, or to hand in homework assignments over the Internet…I'd hate to see this where instructors think, 'Oh I can do everything over the Internet.' It's really impersonal. You know, you come to school to interact with an instructor."

Confidence in the Portal

Security, privacy and reliability were all significant concerns for students and faculty.

"Technology. Does It Work?" The portal must be dependable if it is to be adopted as the tool of choice on campus, especially for students who are worried about deadlines for things like applications and assignments. Myra said, "If there's a certain deadline, how secure is it that the admissions are actually going to get that information? Because I know that sometimes I just like to do it in person, because then I know that it's done. Because there's always that chance for a technical error that 'oh, it didn't go through,' and then I'm screwed."

Privacy and Security

Because students want to access their transcripts and grades online, understandably security and privacy are high on their list of priorities for the portal. John asked, "How secure is this going to be, because I wouldn't want my mom to see my transcripts." Similarly, Annie asked, "How secure is this going to be, because personally, I don't want people to see my standing and my grades." Dr. G. said that privacy is a major concern in her opinion.

"If somebody establishes their own portal page, no one should have access to it. No one should know what you have on it. No other person will have access to it. The portal is only one-way, unless you open a door for them to give you the information. Let me give you an analogy. I can open my front door and go out the front door. But if I want information to come in, there's a slot in the door to get messages in to you. The privacy issue is real legitimate. Administration has the ultimate responsibility for that."

WHAT NEXT?

Lessons Learned

So what exactly will it take to make the portal work? The SDSU leaders we interviewed shared concerns about politics and resources: "The biggest concern is getting all the potential campus entities to cooperate. That would be my number one concern. Number two is actually having the resources to pull it off successfully. Because it has the potential of becoming a very important tool in everyone's life and if it doesn't succeed from day one, it is going to have a bad rap and will be very, very difficult to get it to continue successfully."

One faculty member summed up her strategies for success in the following way: "I think that if it's easy to navigate, if it's dependable, if it's secure, if they're receiving absolute expectations from those around them that they use the tools provided through this portal, then I don't think persistence is going be a big problem."

Here we offer our interpretation of what we've heard from students, faculty, leaders and other universities. SDSU leaders agreed, "Nobody is going to think it's perfect for their specifics, so I think it is a compromise situation." We do not believe that there is a perfect solution, nor do we intend to imply that what works for us will apply directly to other situations. However, other universities may benefit from considering our experiences.

Provide Direction and Leadership

Should leadership in a portal project be top-down, bottom-up or somewhere in between? What has helped other institutions to successfully roll out their portals? We found that there's no easy answer or one single way of approaching this issue. Collaboration was one theme commonly expressed by the SDSU leaders with whom we spoke. "The more that people get involved, the more buy-in." On the other hand, "We've heard of the case where it was edicts from the president…and there were some where there was a group of people who put it together, just did it."

Surprisingly, it was a faculty member who touts a more top-down approach: "It seems incredible to me that we do not have a more well-developed conceptualization of this technology and the way we can integrate it, and there doesn't seem to be a body that's making those decisions…I think the university has to play a big role in deciding what those resources are."

Enlist Help Adapting Existing Systems

When we met with the staff at the University of Tennessee in late Spring 2001, we learned that they took a "student-centric" approach and had two overarching goals for their portal. The first was to auto-enroll students from their student information system into their course management system (CMS) so there would be a course roster for every class, allowing grade submission from the CMS. The second goal was to provide CMS certification for every instructor.

UT admitted that one of the main challenges of implementing their portal was in the implementation of lightweight directory access protocol (LDAP). In an attempt to meet this challenge, Tennessee had their advisory groups assist them in adapting their previous course management system to a portal. Similarly, at SDSU we created an LDAP sub-committee of the *ad hoc* portal committee in order to focus on this critical component in portal implementation. The SDSU Director of SIMS/R said that getting everyone to agree, "on type of authentication, which allows everyone with a single login, is a prime concern. People have to change the way they do business." Following the lead of UT, we will be utilizing the EduPerson object class that includes widely used person attributes in higher education. This standard will be the basis for a common list of person attributes for our institution-wide directory (see http://www.educause.edu/eduperson/).

Segment the Rollout

Our strategy will be to introduce the portal in multiple phases, focusing first on where it can have the most impact. The focus group and administrator input from SDSU has a strong academic focus and so the first round of users will be faculty and students.

"More of an Academic Thing." The data we gathered confirmed our belief, suggesting that the initial focus should be an academic one. For students, this means, "links to Admissions and Records, Financial Aid, registering for classes" all in one place. As Josh said, "The number one priority is getting all the SDSU resources tied together and on first before getting the news and sports."

Leadership agreed on an academic focus that includes the link between course management and student information systems. "The Student Information System is going to be one of the bangs for the buck with Academic Affairs. The other bang for the buck is faculty, improving their ability to communicate with their students, definitely [providing] the transparent link [from Blackboard] to the portal as a whole, [as well as] moving of the data between SIMS/R and Blackboard. For the initial rollout, first priority is Blackboard working with SIMS/R."

Implement Features Most Useful to Users

Leaders, faculty and students agreed that while it would be nice to have many of the advanced features and frills offered by the portal, we must first provide features that will be most useful to users. One leader said, "An important step is the development of… a basic student portal…based on SIMS/R…not a true portal, but it helps students and eventually it will become more effective [as the basis for the campus portal]."

One faculty member advised against trying to be everything to everyone, and suggested instead introducing the portal in pieces. "What we need are tools that can help us do what we do well…If you could do it in a components basis, so that things can be instituted sooner, rather than having to wait for this sort of Microsoft scheme [gestures—the whole world] that 'we're going to be everything to everyone' and not doing anything well."

Likewise, Tennessee is trying to make some of the people happy instead of trying to be everything for everyone. They've chosen to focus on students and faculty as a leverage point, much as we've chosen to do at SDSU.

Communicate Benefits

We advocate clearly defining the purpose and vision by involving stakeholders and leadership, and using marketing strategies to increase awareness. It may also be valuable to highlight dissatisfaction with the *status quo* and articulate the benefits of a portal with both printed and online materials, as well as through presentations, contests, etc.

Put Information in All Communications Materials

For example, as a few students suggested, a useful tactic might be to include information about the portal in letters sent home to students and their parents, or in orientation materials. Gabby admitted, "The letter that we get about when we register…I look forward to that every semester. I would look at that."

Explain How it Relates

While users can already find many potential portal services online, they may not realize that the value of the portal is in collecting all these disparate systems in one place.

Convince People That It Is Secure and Dependable

Security and privacy were high on participants' list of concerns. When we asked Dr. M. about whether she would be interested in using the portal for personal and staff-related use, she said, "It would have to be explained to me what the security of this portal was before I would feel comfortable." When we asked the students what would help make people feel that the site was secure, Cliff suggested that, "you need some sort of description for the non-computer user, not tech talk."

Provide Organizational Support

It is clear that to ensure success, the university must provide the proper funding, equipment and access to data. Furthermore, we must ensure adequate personnel, policies and procedures, and allow users time to get up to speed. One faculty member suggested: "The university has to be committed to putting money into the development, to preparing faculty, staff and students who will use it, and to the setting of expectations that it will be used and that it will take the place of other labor-intensive and therefore fiscally expensive ways of doing things."

Based on user input, we plan to use the following strategies in our implementation:

- *Provide Training and Support:* We will include job aids, resource materials, tech coaches, walk-up help desk and online tech support.
- *Encourage Collaborative Learning:* We will foster user "communities of practice" and informal "peer" learning/coaching.
- *Reward and Recognize Users:* We will highlight intrinsic rewards that come with seeing growth and visible results and provide extrinsic rewards for the users' extra effort (e.g., peer recognition, celebrations of milestones).
- *Plan for Evaluation:* Employ formative and summative data-collection methods that utilize online rating forms, interviews, surveys, focus groups and observation (e.g., data tracking, review of marketing and training materials, usability testing).

FUTURE STEPS

In phase three of the SDSU portal rollout project, we will target the campus-wide population of students and faculty through a modified portal survey. We will use the survey data to help us further refine project scope, helping us decide which features can and should be delivered first, and to determine the initial target audience for the portal (e.g., all SDSU students? first-year students only?).

The SIMS/R group is currently working on online registration, change of address, change of major, unofficial transcript and class schedule features for students. For faculty, the SIMS/R group is working on grade submission, posting of rosters and integration with the Blackboard system. In fact, a portal prototype is being developed using Oracle 9i development tools.

Focusing first on the academic interests of faculty and students, the portal will improve their ability to communicate by providing a transparent link between SIMS/R and Blackboard. For instance, at their fingertips, students will be able to access their current academic activities, including readings, projects, deadlines and exams anytime, anywhere, from around the world. A viable Oracle prototype has given us valuable experience that will be applied to building the campus portal.

CONCLUSION

As described in the article, "Charting a Smooth Course to Portal Development" (Frazee, J.P., 2001), SDSU is relying on data to support participative decision making. In this chapter we described how we documented the voices of key stakeholders through a process aimed at developing a rich description from a variety of perspectives. It is important to point out that this type of descriptive approach (interviews, surveys and focus groups) requires a considerable amount of time in order to accurately describe the context or setting of the project. We paid careful attention to hold true to this approach, and attempted to refrain from direct interpretation. Instead, we used narrative to present the voices of end users. Our final interpretation of the data was informed by, and developed in light of, the literature of change and technology adoption.

San Diego State University is striving to achieve success by beginning with the end-users, not only keeping them in mind, but also involving them in the process. From the *ad hoc* portal committee, with representatives from the various campus stakeholder groups, to the focus groups, surveys and telephone interviews, an effort has been made to involve those who will be the end users.

In this chapter, we described the process we have used so far and hinted at our plans for the future. At SDSU we have worked together to develop a list of critical features, a rubric for judging software products and recommendations on the process to follow. We have also developed procedures for further involving stakeholders through interviews, focus groups and surveys. We have collected data using these instruments, and have categorized our findings. Furthermore, we have begun developing and pilot testing various elements that will be included in our portal.

We realize that each campus has different needs, capabilities, politics and issues that will shape the way in which they proceed to develop their portal. We hope that some of the processes we have used, and lessons we have learned, will be useful to others as they plan for a portal that best serves their end users.

REFERENCES

Asmussen, K. J. & Creswell, J. W. (1995). Campus response to a student gunman. *Journal of Higher Education*, 66, 575-591.

Connolly, C. G. (2000). From website to portal. *EDUCAUSE Quarterly*, 3(2), 38-43.

Covey, S. R. (1989). *The 7 Habits of Highly Effective People*. New York: Simon & Schuster.

Ely, D. P. (1999). Conditions that facilitate the implementation of educational technology innovations. *Educational Technology*, 39(Winter), 98-305.

Frazee, J. P. (2001). Charting a smooth course for portal development. *EDUCASE Quarterly*, 24(3), 42-48.

Frazee, R. V. (2002). Technology adoption: Bringing along the late comers. In Rosett, A. (Ed.), *The ASTD E-Learning Handbook* (pp. 263-277). San Francisco: McGraw-Hill.

Fullan, M. (1999). *Change Forces: The Sequel*. London; New York: Falmer Press.

Gilbert, S. W. (2000). Portal decisions demand collaboration—Can portals support it? *Syllabus* (September), 1-3.

Lincoln, Y. S. & Guba, E. G. (1985). *Naturalistic Inquiry*. Beverly Hills, CA: Sage.

Rossett, A. (1999). *First Things Fast: A Handbook for Performance Analysis*. San Francisco, CA: Jossey-Bass/Pfeiffer.

Stake, R. (1995). *The Art of Case Study Research*. Thousand Oaks, CA: Sage.

Strauss, A. & Corbin, J. (Eds.) (1994). *Grounded Theory Methodology: An Overview*. Thousand Oaks, CA: Sage.

Yin, R. K. (1989). *Case Study Research: Design and Method*. Newbury Park, CA: Sage.

ACKNOWLEDGMENTS

The authors wish to acknowledge the contributions of Susan Mar, Eric Tremmel, Jinpil Shin, Doug Fisher, Michael Reese and Kirsten Hansen. And, to the many students and faculty members who contributed by thoughtfully participating in the interviews, surveys and focus groups, thank you!

ENDNOTES

[1] This two-person team performed in the capacity of consultants to ITS as part of EDTEC 644, "Seminar in Advanced Instructional Design."
[2] EDUCAUSE: a professional organization devoted to IT in higher education.
[3] Benefits matrix categorized potential benefits as convenience, capacity, communication, collaboration or cooperation.
[4] Hall's Stages of Concern include Awareness, Informational, Personal, Management, Consequence, Collaboration and Refocusing.

APPENDIX A: SDSU PORTAL RUBRIC

	Insufficient	Adequate	Excellent	Score
I. Look & Feel (This refers to the front-end itself, not the external resources linked to it.)				
Aesthetics	0 points Static background with few or no graphic elements. No ability for variation in layout or typography.	1 point There are a few graphic elements and there is limited ability for variation in type size, color and layout.	2 points SDSU has full control of look and feel, and changes can be made quickly. Appealing graphic elements are included appropriately. Differences in type size and/or university colors and logos are used well.	
Ease of Use	0 points Counter-intuitive interface, requiring greater than two hours of user training.	1 point Somewhat intuitive interface, requiring two hours or less of user training.	2 points Intuitive interface, requiring little or no user training.	
II. Security				
Authentication	0 points No authentication-lacking digital credentials when user logs-in.	1 point Requires multiple log-ins in order to access different databases-limited digital credentials, e.g., Kerberos.	2 points Single sign-on for multiple functions from one central database. Takes advantage of Web browser-friendly public key certificates.	
Access	0 points Information access is all or nothing.	1 point User is allowed to access certain information based on their user type. There is a limit to the number of roles a user can have in the system.	2 points User is allowed to access and change certain information based on who they are and their user type. There is no limit to the number of roles a user can have in the system.	
Hosting	0 points Server(s) under control of vendor at location undetermined by SDSU.	1 point Server(s) located at SDSU.	2 points Server(s) located at SDSU or at a location approved by SDSU.	
III. Personalization				
Information Push	0 points User receives targeted information relevant to their constituency, e.g., pushed to a senior.	1 point User receives information relevant to the individual, but limited dynamically updated data available.	2 points User receives specific information relevant to the individual and available in real-time. For instance, student-specific course schedule, enrollment details and degree checklist.	
Information Pull (Portal Editor)	0 points No editing tool to customize the portal environment.	1 point Editing tool for customizing tabs, panel buttons colors and fonts. Personalized view of all the information relevant to user-specific needs and preferences.	2 points Editing tool for full customization as well as the ability to create discussion boards, chat, etc. User has ability to add/edit/remove information from a list of internal and external resources that the university approves. Built-in translator supporting multiple languages.	

APPENDIX A (CONTINUED)

Link to Existing Course Management System	0 points No link.	1 point Partial access to Web-enabled classes.	2 points Full interoperability with course management system.	
IV. Interaction				
Email	0 points No email.	1 point Portal only accommodates proprietary email system.	2 points Supports multiple email standards and protocols, e.g., IMAP or POP.	
Chat & Message Boards	0 points No chat or message board functionality.	1 point Only chat or only message board.	2 points Supports real-time chat and message boards.	
Electronic Balloting and Polling	0 points No electronic balloting and polling functionality.	1 point Criteria for balloting and polling are only partially supported.	2 points Electronic balloting and polling are fully supported.	
Multimedia	0 points No streaming audio or video.	1 point Options for streaming audio and video limited.	2 points Support for plug-ins that allow for streaming audio and/or video.	
V. Productivity Tools				
Search Engine	0 points No search engine.	1 point Limited search engine for university intranet and/or Internet only.	2 points Natural language search engine for both intranet and Internet e.g., internal Ask Jeeves.	
Calendar	0 points No calendar.	1 point Shared calendar available.	2 points Personalized calendar is available, utilizes IETF standards and allows others (w/user approval) to populate their calendar. Synchronization with Palm OS is available.	
Meeting Scheduler	0 points Does not support a campus-wide meeting scheduler.	1 point Limited campus-wide meeting scheduler available.	2 points Campus-wide meeting scheduler for specific users with ability to select and reserve specific rooms and equipment.	
To-Do List	0 points No to-do list.	1 point To-do list available, but with limited features.	2 points To-do list available, but with many features, e.g., items can be placed in categories and ranked in priority order.	
Address Book	0 points No address book.	1 point Address book is limited and proprietary.	2 points Address book can interface with other, more popular contact lists and databases.	
VI. eCommerce				
Advertising Control	0 points Banner advertising on every page of portal or not able to control per user group.	1 point Banner advertising is optional.	2 points Banner advertising is optional, controllable and can be targeted to specific user groups based on their role within the university.	
Advertising Revenue	0 points No advertising revenue possible.	1 point Advertising revenue is limited and shared with vendor.	2 points Advertising revenue goes directly to university.	

APPENDIX A (CONTINUED)

Transactions	0 points Cannot be integrated with campus systems offering Web-based transactions.	1 point Can be integrated with campus systems offering Web-based transactions.	2 points Can be integrated with campus systems offering Web-based transactions with option of one "shopping cart" enabling the user to credit the appropriate entity.
VII. Workflow			
Forms Routing	0 points No forms routing.	1 point Forms routing available– paper documents can be replaced with Web-based forms.	2 points Forms routing available – paper documents can be replaced with Web-based forms. In addition, forms tracking software built in.
VIII. Support			
Integration	0 points No fit with existing Open Data Base Connectivity (ODBC) relational databases e.g., Oracle RDBMS.	1 point Some fit with existing ODBC relational databases e.g., Oracle RDBMS.	2 points Great fit with existing ODBC relational databases, e.g., Oracle RDBMS.
Implementation	0 points Vendor relies solely on SDSU staff.	1 point Vendor provides implementation training to SDSU staff and consultants for a fee.	2 points Vendor provides implementation training to SDSU staff and on-site consultants for free.
Maintenance	0 points No plan for ongoing support or maintenance.	1 point Weak plan for ongoing support or maintenance.	2 points Strong plan for ongoing support or maintenance.
24/7 Help	0 points Vendor requires toll call during business hours. No email or Web-based help.	1 point Vendor provides email and Web-based help. No phone or fax help (24/7).	2 points Vendor provides email, Web-based and toll-free help (24/7) for free.
Long-Term Viability	0 points Vendor is in pilot phase and has no experience or references. Small company with limited funding.	1 point Vendor has experience in higher education portal development, but has limited references and some funding.	2 points Vendor has significant higher education portal development experience, can provide numerous references and is part of company with ample financial backing.
IX. Standards			
API (Application Program Interface)	0 points Portal API cannot pass information to other applications, or is not available to campus.	1 point Portal API can pass some information to other applications and is available on a limited basis.	2 points Portal API can pass security information to other applications, seamlessly integrating multiple sources of information and campus can write their own interface, e.g., providing a single sign-on environment.
LDAP (Lightweight Directory Access Protocol)	0 points Portal is not LDAP compliant, e.g.; it will not allow a user to query a database via the Internet.	1 point Portal is LDAP compliant, but only allows limited online querying.	2 points Portal is LDAP compliant and allows user to actively manage and customize a personal database.
ADA (Americans with Disabilities Act)	0 points Vendor makes no accommodations for those with special needs.	1 point Vendor has limited features for those with special needs.	2 points Vendor provides screen reader and other features for those with special needs.

APPENDIX A (CONTINUED)

X. Administration				
Staffing	0 points Seven or more full-time SDSU staff required for managing and maintaining system software. Vendor relies solely on SDSU staff.	1 point Between four and six full time SDSU staff required for managing and maintaining system software. Vendor provides training to SDSU staff and consultants for a fee.	2 points Three or fewer full-time SDSU staff required for managing and maintaining system software. Vendor provides training to SDSU staff and data migration/integration consultants for free.	
User Definition	0 points System will NOT allow SDSU administrator to define custom user types.	1 point System will allow SDSU administrator to define limited user types.	2 points System will allow SDSU administrator to define custom user types.	
Information Channels	0 points System will NOT allow SDSU administrator to define custom information channels.	1 point System will allow SDSU administrator to define limited information channels.	2 points System will allow SDSU administrator to define custom information channels. No limit to number of information channels.	
Time to Market	0 points System will take greater than: 8 weeks to define 12 weeks to design 8 weeks to prototype 16 weeks to rollout	1 point System will take: 8 weeks to define 12 weeks to design 8 weeks to prototype 16 weeks to rollout	2 points System will take less than: 6 weeks to define 9 weeks to design 4 weeks to prototype 14 weeks to rollout	
Hardware Resources	0 points Hardware requirements do not coincide with university standards.	1 point Hardware requirements loosely coincide with university standards.	2 points Hardware requirements coincide with university standards.	
Pricing	0 point Annual license fee and Service Level Agreement costs not based on fixed price schedule or exceeds budget.	1 point Annual license fee and Service Level Agreement costs based on fixed price schedule.	2 points Annual license fee and Service Level Agreement costs based on fixed price schedule and are under budget.	
Online Help, Documentation & Training	0 point No help putting portal applications into production.	1 point Plan for integration, but little documentation and training.	2 points Clear integration, with plenty of supporting documentation and face-to-face train-the-trainer training. Several online help features, e.g., tutorials, job aids and FAQ's.	
Smart Card	0 point No support for smart card technology.	1 point Limited support for smart card technology.	2 points Full support for smart card technology.	

APPENDIX B: SDSU PORTAL STUDENT SURVEY

Question 1: In which college are you enrolled?

Question 2: What year in school are you?

For Questions 3 through 8, using the scale, "Strongly Disagree, Disagree, Agree, Strongly Agree," please indicate how strongly you agree or disagree with the following statements.

Question 3: 'I am able to...' [Internet Use]
    Question 3: 1. Use the Internet more now than two years ago
    Question 3: 2. Do most of the things I need/want to do using the Internet
    Question 3: 3. Use the Internet to communicate and work with others
    Question 3: 4. Use the Internet in ways that contribute to my academic success

Question 4: 'I would like to use the Web to: ____.' [Academic Use]
    Question 4: 1. View course syllabus, assignments and due dates
    Question 4: 2. View detailed grades and class standing
    Question 4: 3. Take exams
    Question 4: 4. Submit papers
    Question 4: 5. Access course materials
    Question 4: 6. Conduct research

Question 5: 'I would like to use the Web to: ____.' [Administrative Use]
    Question 5: 1. Pay SDSU tuition and fees
    Question 5: 2. Apply for financial aid, check for and receive notifications of status
    Question 5: 3. View SDSU courses and course information prior to registration
    Question 5: 4. Register for courses
    Question 5: 5. Order transcripts
    Question 5: 6. Access and update my personal SDSU records
    Question 5: 7. Buy things (e.g., textbooks, event tickets)
    Question 5: 8. Apply for SDSU housing and search for off-campus housing
    Question 5: 9. Check progress toward my degree
    Question 5: 10. Change my major

Question 5: 11. Change address
Question 5: 12. Reserve materials at the library
Question 5: 13. Reserve book return notifications
Question 5: 14. Reserve study rooms at the library
Question 5: 15. Automatically receive bulletins and announcements from my school or college
Question 5: 16. Find directions and/or maps to a variety of campus locations
Question 5: 17. View days and hours of operation for all campus buildings, offices, services and businesses
Question 5: 18. Review course and faculty evaluations

Question 6: 'I would like to use the Web to:____.' [Communication Use]
Question 6: 1. Share files with other instructors, students and others
Question 6: 2. Use streaming video and/or other advanced technology applications
Question 6: 3. Use online discussions and forums
Question 6: 4. Present work (e.g., make research results available online)
Question 6: 5. Work with others on special projects/assignments via email, chat, online calendar, etc.)

Question 7: 'I would like to use the Web to:____.' [Personal Use]
Question 7: 1. Post resumes and view job openings
Question 7: 2. Interview with prospective employers
Question 7: 3. Join clubs, socialize, etc.
Question 7: 4. Automatically receive news from the SDSU newspaper
Question 7: 5. View a variety of campus events by area of interest
Question 7: 6. Schedule an appointment at SDSU Health Services

Question 8: 'I would like to use the Web to:____.' [Miscellaneous Use]
Question 8: 1. Receive automated reminders for assignments, appointments and events
Question 8: 2. Save materials to an online file storage
Question 8: 3. Access saved bookmarks ('favorite' websites) from any computer
Question 8: 4. Access my personal calendars, course schedules, to-do lists and address books from any computer

Question 9: Which three issues most concern you when using the Internet?
        Question 9: 1. Security of electronic data
        Question 9: 2. Privacy of communications
        Question 9: 3. Cost of technology
        Question 9: 4. Reliability of technology
        Question 9: 5. Access to technology
        Question 9: 6. Ethical use of electronic information and technology
        Question 9: 7. The time it takes to learn and use technology
        Question 9: 8. Lack of necessary technical support
        Question 9: 9. Technology standards
        Question 9: 10. Cross-platform problems (Mac-PC-UNIX-LINUX)
        Question 9: 11. Speed of the system
        Question 9: 12. Other

Question 10: How many different passwords do you maintain for SDSU online services?

Question 11: In order of priority, what three things could SDSU do to improve its Internet services?

Chapter X

Values-Based Design of Learning Portals as New Academic Spaces

Katy Campbell and Robert Aucoin
University of Alberta, Canada

ABSTRACT

Many guidelines for portal design tend to focus on the technical aspects of a portal or a network. However, as we continue to define portals as gateways for learning, we need to consider issues related to the social and cultural context in which portals are used. In this chapter we examine learning portals from both the instructors' and the learners' perspectives by synthesizing existing research and proposing a framework for quality guidelines.

The Collaborative of Online Higher Education Research (COHERE), consisting of eight large research-intensive universities in Canada involved in Internet-based learning, was created to enhance learning and teaching through technology and to move toward a stronger culture of professional collaboration and scholarship in our educational practices (Carey, 2000). Based on our experience with COHERE, we have developed tools for the formative and summative evaluations of learning portals generally. These tools include usability studies, questionnaires and focus groups.

According to Boettcher and Strauss (2000), the portal concept dates from the advent of Internet Service Providers (ISPs) like Prodigy and AOL and, later, search engines and interfaces such as Yahoo! and Netscape. Since then the concept of information portals has expanded to include consumer portals, community portals, corporate portals and vertical portals, all of which provide a more customized information experience (Looney & Lyman, 2000). That is, the interaction of the user with the portal's information offerings can be personalized based on previous and current user choices which, taken together, form a dynamic user profile. This attribute of portals, among many others, exemplifies the potential for portals as learning environments.

Portals have been described as "a single integrated point for useful and comprehensive access" (Eisler, 2001); "a new umbrella Webpage array...that encompasses many or all of the current homepages for departments and individuals" (Batson, 2000); "an internal consolidation of online services to be provided via the Web for faculty, staff and students" (University of Montana); "an integrated platform that lets people interact in real-time with a company's systems and information" (Copeland, 2001); "(having) the capability to aggregate content and integrate workflow from multiple sources, access role-based analytical information and facilitate transaction" (Norman, 1999); a hub (Boettcher & Strauss, 2000) and as a user-created, one-stop Webpage of collected information (Looney & Lyman, 2000).

These are functional definitions of a portal as an integrated system providing a gateway to organized data. They hold in common a set of functions and outcomes that enhance and democratize access to information. However, a learning portal may go beyond the information management function to create new learning communities and academic spaces that enable *profoundly redefined relationships* among teachers, learners, and the institution and its external communities. Portals provide important mechanisms for reaching out to new populations of learners and engaging them in new ways to facilitate learning and development. Beyond serving as a gateway and an organizer, a portal can provide access to a broader range of contemporary information and learning resources (experts, teachers, researchers, mentors), encourage enriched interaction with those resources and with other learners, wherever they may be in the world, and support new models of teaching, learning and research.

In this chapter we attempt to describe these new spaces and relationships in a context of cultural change in higher education. We discuss the common attributes of a well-designed portal and, using evaluation criteria developed by research in human-computer interfaces and related fields, we suggest components of a framework for portal design and evaluation that empower faculty and learners to be both participatory designers and critical users of portals. Finally we ask, "What institutional issues must be considered in the collaborative design of a learning and/or research portal?"

PORTALS AND A TRANSFORMED LEARNING ENVIRONMENT
Emerging Challenges for Post-Secondary Institutions

As post-secondary institutions come to terms with the "new knowledge economy," they must acknowledge the forces and sources of change that have and will affect the shape of higher education (HE) for the next decade. As a result of an institution-wide four-year strategic planning process, our faculty produced a report, *Reaching Beyond*, in which we have identified six forces that relate to demographic and sociopolitical issues that inform the design of learning environments and communities in HE: the changing nature of the learner, the global information society, access and the digital divide, the increased emphasis on the "business of education," the changing nature of work and the emergence of the consumer culture in education. These factors suggest a renewal, and perhaps a profound reorientation, of the traditional post-secondary institution.

The demand is growing for institutions to provide flexible programs and points of access to the learning environment for learners from increasingly diverse backgrounds. We must be able to relate to the needs of people from a wide range of age cohorts, gender differences, ethnic and cultural contexts, family mobility and changes within the workforce.

The emergence of new and evolving occupations that are expert-defined, interdisciplinary in character and not encompassed by traditional university structures is a significant trend. Professional associations will demand transparency in quality assurance of practices that grant credentials. Lifelong learning is becoming a fundamental source of employment security in an age of rapid change and globalization. At the same time, in an effort to find alternative sources of funds for higher education, corporate sponsorships and partnerships are increasingly sought. Some of these initiatives have and will be focused on the integration and use of information and communication technology in education (see, for example, UNext.com). The resultant cultural shift has had a ripple effect throughout the institution as faculties and departments look for ways to adapt.

Finally, students, as consumers, want to invest in an education that will help to ensure their employability. They are seeking practical knowledge, a technical skill set and credentials that will increase their marketability at various times in their lives. Such students may view themselves as clients who are purchasing the commodity of education, and may well demand accountability for expenses in order to justify participation in their selected programs. Paradoxically, universities must take care that they not merely become training grounds for future employers. Universities must maintain their autonomy. What is important is that learners have a degree of control over their learning, that they learn how to learn and that they learn how to be critical consumers of education.

Universities are seeking ways to manage and facilitate emerging areas of research and discipline specialization, diverse life circumstances and learner profiles, and partnerships with internal and external communities that challenge the autonomy of the single-source institution. Public leaders have expressed a strong interest in alternative methods for delivering, supporting and facilitating learning—any time, any place, any pace—as required in new knowledge-intensive environments and as enabled by converging information and communication technologies. Therefore the decision to implement a campus portal for enhanced learning opportunities must address issues of equity and access, flexibility, innovation, personalization, credibility, quality, transparency and transferability within the framework of evolving institutional goals and strategies.

An examination of the popular and academic literature, and a review of a range of portals on the Internet reveal a set of attributes or features that characterize most commercial portals. In usability language, these are user-defined guidelines that guide the design or acquisition of a portal system, and may be considered as evaluation criteria. Portal Functionality organizes these features by user function (see Figure 1).

Both Campbell (2001) and Batson (2000) contend that commercial portals are built on different values and assumptions than those of the academic community and are seen as pursuing different goals and purposes. Campbell, in particular,

Figure 1. Portal Functionality (Adapted from Eisler, 2000; Boettcher & Strauss, 2000; Paadre & King 2000; University of Montana)

	Tools	Access	Resources	Engines
Features	Internet	single log-in	content	search
	personalization	authentication	library	navigation
	customization	security	support	
	communication	directory	training	
	interaction	gateway to interconnected resources and services	people	
	workflow integration			
	application integration			
	e-business			
	intelligent agents			
	learning management			

discusses the need for a "scholar's portal" that meets the needs of the research community. Believing that the process of portal development may encourage or reflect behavioral changes in an institution, Erhmann (2000) speculates that institutional goals may include service provision; flexibility and responsiveness of instruction; the enrichment and extension of academic communities; attracting and retaining students and staff; fostering universal, frequent use of computing communications; and sustainability. Instructional goals, however, may include changes in behavior among faculty, learners and support staff. Who are the stakeholders of an evolving academic community? How does the design of learning portals align with the various forms and requirements of distributed learning environments? And, how can a portal support a transformed learning environment?

New Spaces, New Partners, New Goals

Batson (2000) suggests that a learning portal expands on traditional academic space, which has traditionally been defined as physical infrastructure--with related resource structures--that shapes the nature of the interactions that occur within it. This traditional space has an important socialization function: members of the community know how to speak and act within these spaces, understand power relationships by the way these spaces organize interactions and, once acculturated, can subvert the purposes of these spaces. The nature of teaching and learning has been entirely defined by a familiar landscape, where learning events were structured by place and time and format.

For all the factors discussed above, that landscape has fundamentally changed. Learners, from undergraduates to professionals, and non-formal learners become more heterogeneous all the time and increasingly demand customized learning experiences that are flexible, authentic and relevant. Faculty, who have old maps with which to navigate this new landscape, must nevertheless redefine their relationships with learners, with new forms of knowledge representation, with research, and with external communities that are suddenly present in their "classrooms" and that are influencing their planning. A constituency that has no brand-loyalty and that expects program mobility challenges administrators who have been operating with a management strategy that focuses on internal factors to the exclusion of the external realities of a "new economy."

In responding to these challenges, HE has invested so heavily in Internet-based technology that the Web is rapidly becoming the software model or learning template for universities and colleges. Although institutions have ranged themselves along an academic space continuum from primarily face-to-face to primarily virtual, most have settled on a technology-enhanced, or distributed approach to learning and access. Employing alternative forms of instructional and delivery models, this approach includes synchronous tools and environments such as classroom lectures, audio and videoconferencing, and data conferencing; and asynchronous tools such as computer-mediated conferencing and other communications systems, learning management systems, and print and digital media. Much of the learning content and interactions can be stored, extended and reused in digital repositories. New ways of supporting new learning communities are available. This approach fundamentally realigns and redefines institutional infrastructure: it is more learner-centric and open in design and support, including extended information services; and has a significant social effect on the academic community, raising fundamental questions about academic freedom; intellectual property rights management; and the nature of knowledge discovery, representation and stewardship.

To "play a new and expanded role in the ongoing education of citizens," a *cultural shift* in universities and colleges is required (Advisory Committee for Online Learning, 2001, p. 24). In rethinking the learning enterprise, HE is acknowledging that new academic spaces change their relationships to the society and the economy as a whole. A cultural shift depends on addressing the concerns of faculty members and involving them in a process of change and transformation in the ways they plan, teach and interact with learners (Bates, 2000), and in developing new, global collaborations with learning partners, including those in a position to provide learning services, solutions and systems.

Learning portals can provide the functionality of consumer systems, but at the same time support the social, cultural and political goals of HE. While resisting, to a greater or lesser extent, the culture of the corporation, universities nevertheless have begun to adopt the concept of portals as learning storefronts (Galant, 2000). Yet, in order to respect HE values of knowledge creation and dissemination for the greater good, these portals must go beyond the functional requirements and gateway view of commercial portals, and exist as tools that both *transform* the academic environment and *represent* it to the world.

In order to define the roles and requirements of learning portals, the stakeholder communities must identify their tasks, roles and principles. In an institution that is in the process of realigning its strategic goals, the identification of tasks and roles may emerge from these principles and values. Gilbert (2000) and Eisler (2000) identify major categories into which a variety of features and functions can be organized: gateways to information, points of access for constituent groups and community/learning hubs. A synthesis of public reports suggests the stakeholders and their functional requirements shown in Figure 2.

How this incomplete list of requirements can be embedded in the context of a transformed learning environment is the focus of the remainder of this chapter.

Figure 2. Needs of Stakeholders (Adapted from the work of many institutions that have shared their experiences over the Web, including the University of Southern Florida, Purdue University, Holy Cross, University of Montana and Princeton)

Stakeholders	Teaching/Learning/Research		Administrative Systems				External Communities			Learning Systems	
Features	Faculty	Learners	Mgt.	Financial	Staff	Facilities	Alumni & Visitors	Parents	Partners	Library	Network
Track applications	✓	✓			✓			✓	✓		
News & events	✓	✓			✓		✓	✓	✓	✓	
Advising	✓	✓	✓		✓		✓	✓	✓	✓	
Registrations	✓	✓	✓		✓		✓	✓	✓	✓	
Fines & Fees	✓	✓		✓	✓	✓					
Records	✓	✓	✓	✓	✓	✓	✓	✓	✓	✓	✓
Scheduling	✓	✓			✓	✓					
Announcements	✓	✓	✓	✓	✓	✓	✓	✓	✓	✓	✓
Group discussions	✓	✓			✓		✓	✓	✓		
Messaging	✓	✓	✓	✓	✓		✓	✓	✓	✓	✓
Transfer	✓	✓	✓	✓	✓		✓		✓	✓	
Class lists	✓	✓			✓	✓				✓	
Course development tools	✓	✓			✓				✓	✓	
LMS	✓	✓	✓	✓	✓	✓			✓	✓	✓
Research accounting	✓		✓	✓	✓				✓		
Operating budget	✓		✓	✓	✓	✓			✓	✓	
PD	✓	✓			✓		✓	✓			
Planning											
Access to resources	✓	✓	✓	✓	✓	✓	✓	✓	✓	✓	✓
Assessment	✓	✓	✓					✓	✓		

VALUES-BASED PORTAL DESIGN

In the previous section we suggest that campus learning portals are integral to the cultural shift in HE as they encourage transformational thinking about relationships and spaces, and represent that shift to the external (and internal) world.

The values upon which this new learning environment might be based include:

- *Inclusiveness:* The portal design must support diverse communities, including the older professional; the distance learner; the non-traditional learner; the physically challenged learner; the workplace learner; the learner with alternative language, cultural and perceptual needs; both present and virtual faculty; multidisciplinary teams of researchers; local and international academic, business and political partners; and others.
- *Integration:* Learning management systems such as Blackboard™ and WebCT™ originally offered a publishing and course management environment with an integrated tool set for faculty, including tools for communication, assessment and record keeping. These companies, among others, have begun to develop and refine enterprise systems, which integrate instructional, delivery and administrative systems in the institution. These portals have evolved from a teaching/learning orientation and reflect institutional movement towards a seamless, integrated learning environment that meets the needs of many constituencies.
- *Learner-centeredness:* Traditional institutional websites have been very owner-centric. Portals, both by definition and by design, are user-centric. Portal design is based on the interrelated concepts of customization and personalization. This orientation reflects more the perspective of the learner (or education consumer) in learning environments, in which learners can build learning portfolios based on their circumstances, experiences and current needs.
- *Accessibility:* Pressures of the new economy imply that the intellectual resources of the university should be packaged and made available to a global community. Portals identify, organize and represent these resources in ways that make them easy to retrieve, use and reuse (see, for example, MIT's Open Courseware Initiative).
- *Flexibility:* Customizing and personalizing learning experiences addresses the dynamic needs of the lifelong community. For many reasons, including changes in professional accreditation, a globally mobile workforce, new and emerging occupations and life events, individuals in this community will search for opportunities to time-shift, place-shift and construct individual programs from many providers. A well-designed learning portal will act as a gateway to these opportunities (see, for example, Fathom.com). As more resources are

included, a well-designed portal will be scalable, encompassing more resources and providing access to meet the needs of new communities.
- *Transparency:* A learning portal makes the institution's strategic directions visible to the community. The institution's partners—the learners, external research communities, the private sector and others—construct their own "footprint" to search for all the services they need and deal directly with the systems that facilitate their interactions with the environment. Portals can help the community discover and promulgate best practices.
- *Accountability:* As the learning and support environment becomes more transparent, and as learning opportunities become more available and flexible, community members will expect to be able to evaluate the services and resources to which they have access. As rich information hubs, learning portals can make the institution's quality framework apparent and available for querying.
- *Expanded and blended learning communities:* A learning portal manages transparent and reliable communication tools, which increase access to resources and social learning communities. These tools are easily accessible from the portal and can therefore include and support group members from different institutions, organizations, regions, non-formal communities and cultures. These communities broaden and enrich the learning environment and enhance inclusiveness. At the same time, universities remain concerned at the degree of mobility and flexibility their "customers" are beginning to demand. Looney and Lyman (2000) believe that the value of a learning portal is that "it can be used to engage constituent groups, *empower* them with access to information resources and communication tools, and ultimately *retain* them by providing a more encompassing sense of membership in an academic community" (p. 33).
- *Evened-out hierarchies:* Learning portals have the potential to flatten organizational structures that were inaccessible, accepted non-critically and even unknowable before. With the organization's physical clues missing, virtual academic spaces do not support status clues to the same extent as traditional spaces. For example, a typical classroom, with desks in rows and with a lecture podium at the front of the room, provides social clues about how to behave in this space. Learners expect to learn individually, to receive information from the expert at the front of the room, to take notes and to be cued to leave for the next class at the sound of a bell. Their relationship to the instructor is signaled by his/her relative position in the room, tone and potentially by their appearance (e.g. older, dressed more formally, etc.). Implications include a democratization of "transactions" within and external to the institution.

- *Collaboration:* A campus learning portal is going to fundamentally change the way universities treat its intellectual capital, increasing opportunities for collaborative work on campus, nationally and internationally. It is critical to involve the owners of this capital in the design of the portal environment: faculty members, support staff, librarians, learners, administrators, alumni, the public, and partner institutions and communities. As these constituencies engage each other and become participatory designers, a deeper, transformed understanding of the whole knowledge management enterprise will emerge.

Based on the foregoing values and goals, a picture of a responsive campus portal emerges. The new environment could be designed to include a wide-range of information, communication and development tools. These tools could be divided into categories such as tools for learning environments, tools for research and tools for administrative support. Examples of such tools could include some of the following:

Resources and Services for Instructors and Students
- an array of interactive multi-media tools
- extended elements of traditional library services
- increasingly rich interlinked libraries of both traditional and electronic resources
- access to extra-curricular virtual events
- a high-value, broadband-enhanced learning object management system and repository

Learning Environments and Tools
- intelligent agents, such as "tutors"
- electronic course space
- access to learning support systems
- interactive discussion spaces, open to the world
- an integrated suite of tools for instructional designers, content authors, instructors and learners
- flexible delivery platforms

Research and Administrative Support
- publishing tools
- links to the student information system
- a capacity for individual users to customize and organize personal resources
- resources external to the campus
- horizontal links among departments on campus and vertical links to national academic fields

(Adapted from design guidelines proposed by Eisler (2000), Erhmann (2000), Paadre & King (2000), and others).

EVALUATING A LEARNING PORTAL

Batson (2000) makes the point that we are all co-researchers as we collaborate in the analysis, design and implementation of campus learning portals. Alan Cooper (2001), a regular contributor to *The ZDNet Developer*, agrees that good design is founded on a deep understanding of both broad human characteristics and the specific intentions of a particular constituency or constituencies. A well-designed portal will enable them to achieve both personal and professional goals. As we bring our different roles and goals to the table, we can use the criteria established in other fields (e.g., usability) to guide our planning. Related guidelines include technical criteria; economic/business case criteria (cost-benefit analysis); and criteria extrapolated from research in the areas of human computer interface (HCI), multimedia/hypermedia design, information literacy and learning effectiveness research. As we are most concerned with the social and cultural contexts and implications of learning portals, we have concentrated on the latter group.

Usability Guidelines from HCI

Usability refers to the relationship between tools and their users. An effective tool, website or system must allow intended users to accomplish their tasks most effectively. Usability is the quality of a system that makes it easy to learn, easy to use, easy to remember, error tolerant and subjectively pleasing.

Usability depends on a number of factors, including how well the functionality fits user needs, how well the workflow meets user goals and how responsive the application is to user expectations. Using learning design principles and design guidelines in interface design involves a process of getting information from people who actually use the system—the developers, the learners and management (from *Usability First*).

Numerous authors describe factors affecting usability. The list offered by Frontend.com provides a useful summary:
- *Motivation:* Users prefer to use a service that is useful, relevant and easy to use.
- *Internationalization:* Interfaces must be adaptable to the needs of diverse cultures. An internationalization process should ensure that the interface design can support alternative means of presenting content.
- *Long transaction times:* Users should be able to jump in and out of their learning "transaction" without losing their progress.

- *Relating learning simulations to real-world experiences:* Increasingly used as experience-based e-learning tools, simulations require an understanding of how online students experience their environments in order to teach them on those terms.
- *Device independence and usage contexts:* Desktop PCs, mobile phones, personal digital assistants (PDAs), digital TV and gaming consoles are increasingly used to access the Internet. People will want to interact on a variety of devices and in a variety of places.
- *Meeting the needs of diverse user groups:* It is important to develop interfaces that are flexible enough to support users from a variety of cultural, linguistic and sociological backgrounds.
- *Up-front information:* People need to know what will be required of them before they commit to a specific task.
- *Feedback:* Feedback is essential to help users recover from errors and to assure them that the system is working.

A subset of usability research focuses on accessibility on the Web. Some estimate that up to 90% of all websites present barriers to users with permanent or temporary physical or cognitive disabilities. Jacob Nielsen believes that with current Web design practices, users without disabilities experience three times higher usability than users who are blind or who have low vision. However, principles of *accessible* Web design are usually the principles of good design. A checklist of accessibility design guidelines can be found on the *W3C Web Accessibility* site (http://www.w3.org/TR/WCAG20).

Finally, usability criteria address the cultural relevance or inclusivity of a site. Ito and Nakakoji (1996) reject mere translation of online resources and services as a superficial solution to the challenge of internationalization, because that solution is at the level of technical and national localization. Instead, they plead for "*cultural* localization, which means dealing with values, tastes and the history of the user's culture by going *beyond* surface-level adjustment" (p. 121, the emphasis is ours).

Hypermedia Design

Hypermedia environments support self-directed, lifelong learning if structured to stimulate and motivate learners to be able to independently locate the resources necessary to continue learning (Diaz, 1998). Marchionini (1992) stresses this point: the user wants to achieve his/her goals with the minimum of cognitive load and the maximum of enjoyment. Research shows that users are often unable to explore hypertext without experiencing navigational problems at some point. Detriments, other than the failure to provide an adequate overview of the scope of resources, include cognitive overload, inefficiency because more time is spent learning how to navigate rather than processing information and interference with the critical and creative comprehension necessary to solve open-ended problems (Oliver, 1999).

Hypermedia supports diverse learning and cognitive styles through multiple presentations of information (cf., Daugherty & Funke, 1998). As well, a social environment results in learning gains and increased creativity of outcomes that develop from collaborating and working in groups (Nelson & South, 1999). Internet-based communication tools such as e-mail, threaded discussion forums and synchronous conferencing enable dialogue that can help students think critically and make better decisions. Interaction, especially in cooperative learning activities, appears to be a key factor for success in many hypertext-based learning tasks. Equally important, however, is the intellectual and technical support provided as users learn to navigate these environments and structure their own learning in ways appropriate to the learning tasks and outcomes.

The essential components of an effective hypermedia environment relevant to learning portals, then, are: well-defined goals and explicit scaffolding support such as those provided by intelligent agents and coaches (cf., Guzdial & Kehoe, 1998); authentic learning environments in which knowledge is socially constructed in formal and informal shared spaces (cf., Denning & Smith, 1998); multiple representations of content (cf., Gillham & Buckner, 1997); navigational/cognitive devices such as spatial and conceptual maps and tutors; the selective use of outsiders or virtual guests, for complementary insights and information; and collaboration (cf., McLellan, 1997).

Information Literacy

Over the past decade the academic library community has expanded the tenets of information literacy to include the critical evaluation of Web-based resources. Assessment criteria include site accessibility, architecture and navigation, and interface design as well as content-related issues (cf., Bakken & Armstrong, 2000; Everhart, 1996; Grassian, 1995).

Site accessibility refers to both technical and social access to information. For example, the information should be available in more than one format, especially if plug-ins, increased bandwidth or other tools are required for viewing; multimedia elements should be integral rather than superfluous; graphical elements should respect cultural diversity; and the perceptual, physical and cognitive challenges of a diverse user population should be addressed.

Site architecture and navigation should align with the goals of the site, represent a clear relationship of ideas and concepts, and be of appropriate depth. In an accessible site, the user can easily move around, locate relevant information and receive assistance when requested.

The user should not have to expend more cognitive effort on the interface than on the learning or information retrieval task; the interface must be transparent, with appropriate and meaningful use of colors, icons, menus, etc.; and the site interface must support the site's and/or institution's mission.

In terms of content, many guidelines include questions about author authority, credibility, reliability, availability and integrity. The balance between internal and external links is significant; the links should be current and functional. Information should be current, accurate and relevant; free of bias or at least transparent in perspective; and contain clear, coherent and error-free language.

Academic library evaluation criteria speak primarily to the needs of the experienced and the novice "research community," and usability issues tend to reflect the ease with which information resources can be identified and analyzed for goodness of fit to the user's academic tasks.

Learning Effectiveness

In the past decade, several initiatives have been undertaken to establish standards by which instructional technology innovations can be evaluated. These initiatives differ in scope and depth, each producing a unique set of criteria for varying phases in the development and use of instructional technology innovations. For example, Reeves and Reeves (1997) propose an evaluative model which involves the evaluation of "ten dimensions of interactive learning on the World Wide Web, including pedagogical philosophy, learning theory, goal orientation, task orientation, source of motivation, teacher role, metacognitive support, collaborative learning, cultural sensitivity and structural flexibility" (p. 59). Ragan (1999) outlines evaluative categories specific to learning goals and content presentation, student and teacher interactions and community building, student self-assessment and course assessment, selection of instructional media and tools, and the provision of learner support systems services.

In 1997, the Western Interstate Commission for Higher Education (WICHE) produced a comprehensive report entitled *Good Practices in Distance Education*. This report examines seven facets affecting distance education, including curriculum and instruction, institutional support, and evaluation and assessment. Chickering and Erhman (1996) have also suggested principles of good practice in undergraduate education as a guide for the use of instructional technologies.

In 2000 the American Federation of Teachers produced a comprehensive report entitled *Distance Education: Guidelines for Good Practice*. Guidelines are provided for fifteen different areas related to the delivery of distance education ranging from the role of faculty in online course creation and delivery to student assessment and achievement to the advertisement of online courses. This is by no means a comprehensive list of all of the evaluative initiatives that have been taken or are currently underway, but serves to illustrate the wide scope of available criteria with which to evaluate instructional technology innovations.

Similarly, the Institute for Higher Education Policy in Washington, DC developed seven quality benchmarks for the institutional evaluation of online programs and services: institutional support, course development, teaching/learning, course structure, faculty support, student support, and evaluation and assessment.

A number of consortia of academic service providers have developed guidelines for accreditation of courses and programs on the Web. For example, the Open and Distance Learning Quality Council (ODLQC) accredits open and distance learning providers in the UK. Accreditation includes a rigorous assessment of a college's administrative and tutorial methods, educational materials, and publicity. All courses and programs are evaluated by criteria established for course objectives and outcomes, course contents, publicity and recruitment strategies, admission procedures, learner support, and welfare, presence of open learning centers, and the reputation of the provider.

A Synthesis

This overview of the evaluation criteria from various disciplines exploring online strategies and resources suggests an integrated framework for institutional decision-making related to learning portal implementation. As a starting point, we propose an overall design matrix, reflecting support for six themes related to institutional change and transformation: supporting new learning communities, redefining teaching/learning relationships, internationalization, collaborative relationships, broadening access, and enhancing quality.

INSTITUTIONAL ISSUES IN THE COLLABORATIVE DESIGN OF A LEARNING PORTAL

In this section, we offer a very brief overview of a collaborative project-in-progress and relate a formative evaluation of an emerging portal design to values in HE proposed in Values Based Portal Design.

In September 2000 a memorandum of understanding was signed by five of the larger Canadian research universities to explore ways that they might collaborate on issues in online learning. The resulting alliance, the Collaboration for Online Higher Education and Research, or COHERE (http://www.cohere.ca) now involves eight research-intensive institutions including Simon Fraser University, the University of Alberta, the University of Calgary, the University of Saskatchewan,

Figure 3. Aligning Principles of Learning Portal Design with Transformational Goals in HE

Design features and principles		Supporting new learning communities	Redefining teaching/ learning relationships	Internationalization	Collaborative relationships	Broadening access	Enhancing quality
Tools	course development	✓					✓
	course management	✓					
	assessment	✓					✓
	accounting						
	communications	✓	✓				✓
	e-business			✓			
	search	✓		✓	✓	✓	✓
	navigation	✓				✓	✓
	financial			✓		✓	✓
Systems	registration	✓					✓
	HR						✓
	facilities management	✓	✓			✓	✓
	information	✓		✓	✓	✓	✓
	recruitment and retention	✓	✓	✓	✓		✓
	cost-sharing			✓	✓		✓
	rights management			✓	✓		✓
Resources	knowledge management	✓	✓	✓	✓		✓
	curriculum exchange	✓	✓	✓	✓	✓	✓
	digital repositories			✓	✓	✓	✓
	extended library	✓		✓	✓	✓	✓
	experts	✓	✓	✓		✓	✓
	colleagues	✓	✓	✓	✓	✓	✓
	advising	✓	✓	✓			✓
	collaborative conversations	✓	✓	✓	✓	✓	✓
Principles	accessibility	✓		✓	✓	✓	✓
	integration of services	✓		✓	✓		✓
	flexibility	✓	✓	✓		✓	✓
	relevance	✓		✓		✓	✓
	reliability	✓		✓		✓	✓
	quality	✓	✓		✓	✓	✓
	transparency	✓	✓		✓	✓	✓
	cultural inclusiveness	✓		✓		✓	✓
	efficiency				✓		✓
	sustainability	✓		✓	✓	✓	✓
	scalability	✓		✓	✓	✓	✓
	customizability	✓	✓	✓	✓	✓	✓
	personalizability	✓	✓	✓	✓	✓	✓
	internationalization	✓	✓	✓	✓	✓	✓
	responsiveness	✓			✓	✓	✓
	security			✓		✓	✓
	support for strategic alliances						✓

(Columns grouped under **Transformational Goals**)

York University, the University of Guelph, the University of Waterloo and Dalhousie University. This collaborative process reflects the goals and values of eight diverse institutions with the common mission of developing an inclusive and integrative research and learning portal. This process was, and is, intended to encourage the partners to work together in new ways to open new possibilities to their learners, reflecting individual institutional cultures, supporting values held in common and encouraging the emergence of a new learning culture.

COHERE—A Portal in Progress

The goal of COHERE is to create an alliance of leading Canadian research universities, working together to improve access to online learning opportunities and to integrate online learning with HE research culture and values (Carey, 2000). More specifically, COHERE is meant to be a learning community which provides a framework for collaboration in online learning on both the teaching and research sides of higher education. Since its inception COHERE has undertaken two major initiatives, described below. The collaborative *process*, itself a key product, is perhaps one of the most important "deliverables" in the COHERE project, of which the evolving portal is an artifact. In other words, engagement in the process is a catalyst for the cultural shifts that may occur in the policies and practices of the partners in the alliance.

In a *Memorandum of Understanding* (2000) signed by the Academic Vice-Presidents of the participating institutions, and later included in a proposal for funding made to the Canadian federal department Industry Canada, COHERE described the teaching initiative as a process resulting in guidelines for administrative processes to allow departments, colleges and program committees to integrate courses from other partners into their online programs; shared educational principles and evaluation methods that reflect the research values of member universities; and collaborative planning and development of future online courses/programs/learning objects, based on the above.

Concurrently, the research initiative established a framework for integrating research into teaching, and supported institutional change technology (c.f., Archer, Garrison & Anderson, 1999). COHERE presented enhanced learning as a "disruptive technology," threatening established traditions and values, and proposed to develop evaluative interventions to involve faculty as proactive agents supporting the change process.

In creating COHERE we quickly realized that we would need some way to present a comprehensive and "COHEREnt" face to the world—a face that represented the change agenda of the alliance. COHERE needed a "brand," so that users would recognize it as a unified organization distinct from similar organizations, while retaining the autonomy of the COHERE members. For our purposes a learning portal was an obvious solution, assuming we designed it with the needs of its users (students, potential students, faculty, administrators, researchers and other partners) in mind. COHERE needed to create a system that could provide an infrastructure or scaffolding on which to build the relationships and deliverables that COHERE would create. To create such a system that could reconcile the goals and aspirations of COHERE member institutions, while preserving the integrity of COHERE as a research alliance, seemed impossible until the idea of a portal was suggested. The COHERE portal had to satisfy four requirements. It had to:

1. *Be dynamic:* The portal had to be dynamic and engaging, not only for the users but because it had to reflect multiple institutions. Since we would have visitors from a variety of contexts and because a visitor's needs and interests change over time the learning portal needed to be changeable over time.
2. *Be customizable:* The COHERE learning portal is a tool more for the individual than the university. As such the user needed more control over the information presented therein than a Website would normally provide.
3. *Enable the individual:* The COHERE portal needed to contain tools and information to help the users, including tools for users to provide their own content. This could include information about specific courses that they may have taken, a personalized selection of discussion groups or a personal calendar.
4. *Have a single point of entry:* Users of the COHERE portal should only require one login to the system in order to access whatever information they needed or wanted.

The COHERE portal actually has two parts: the main, central access point (http://www.cohere.ca) and a second part aimed primarily at students and prospective students (http://www.universityonline.ca). The *university online* section is an interactive database where prospective students can search for courses and programs offered online by all COHERE members. Results of the searches provide the users with links directly to the courses or programs of their choice. The value of this method of displaying the results is that it solves the problem of how to present a coherent look to the portal without violating the integrity of the member institutions. In time we hope that this site will expand to include more information, the ability to register online, the ability to pay fees online, and the ability to examine one's academic and accounting records online. We also hope to create learning "spaces" where users can provide their own content through discussion groups and personal calendars.

The cohere.ca portion of the portal is aimed primarily at university faculty members and administrators who are interested in a more in-depth approach to working and learning online. The portal includes information on how to teach online and how to conduct research online. In the future we envision that this will expand to include online courses on how to teach online, current information on copyright and intellectual property as it pertains to online teaching and learning, and information on issues pertaining to cross-institutional accreditation. In addition we will create learning communities in which researchers and administrators can contribute to each other's learning and development through online discussion groups.

Clearly the COHERE portal does not subscribe to all the properties of a standard portal. It is for this very reason that we have chosen to use COHERE as an example of how education portals can be formatively evaluated. Portals, by their very nature, are dynamic. Therefore there will never be a time when the process is finished. The framework we have outlined is best utilized formatively, yielding information to aid in redesign, instead of summatively.

Earlier in this chapter we identified ten core values we believe need to be addressed in any educational portal: inclusiveness, integration, learner-centeredness, accessibility, flexibility, transparency, accountability, extended learning communities, flattened hierarchies and collaboration. As it evolves, the COHERE portal addresses all of these values to varying degrees as demonstrated in Figure 4.

CONCLUSION

Based on our experience developing educational portals, we believe that the key to success lies in creating a tool that goes beyond "technical" usability requirements. An educational portal that does not support institutional value systems will not be successful. We discovered early in the COHERE consortium that it matters less that the portal subscribes to rigid sets of criteria than it does to make the portal design dynamic. In the case of COHERE, it was paramount that users be able to use the portal to fulfill (at least in part) their personal learning goals and that the members of the alliance were able to "sell" the concept to senior administration as one that would preserve existing values while extending the institutional missions and initiatives in a global marketplace. In this chapter we did not intend to provide a specific prescription of how to build the perfect learning portal. Rather we hoped to provide the reader with a values-based framework that can be used to determine the potential effects of an educational portal to help users achieve learning goals in an environment whose values and actions support those goals. We believe that the collaborative design process encourages participants to surface and examine those goals, and that the design process itself is the catalyst in the cultural shift required of higher education.

Figure 4. COHERE Alignment with Key Values

COHERE Portal Values		
Values	*Achieved Goals*	*Future Goals and Suggestions for Improvement*
Inclusiveness	• Site design allows for access with minimal experience in online learning • Site is uncluttered and can be read (with some difficulty) by screen readers • Site uses no plug ins or multiple media	• The site failed the CAST Bobby approval system for visually impaired users; this will be addressed in the next version • Fonts need to be enlarged
Integration	• Site provides direct links to individual university programs using their own course management programs where available	Site could include the following: • The ability to register online • The ability to pay fees online • The ability to examine one's academic and accounting records online
Learner-Centeredness	• Users can immediately use the portal to search for programs & courses by institution, by subject or by discipline or they can do keyword searches	• As with integration, students want "one-stop shopping": they want to register in one place, pay in one place and have a unified accounting & transcript system; so far COHERE does not address these issues
Accessibility	• Users have immediate access to a wide range of programs at 7 of Canada's largest and best-known research universities. In the short term we expect procedures for cross-accreditation to be in place to make it easier for students to achieve their learning goals	• This process should be even more open and learner centered so that students can identify their learning goals and satisfy them through a single point of entry
Flexibility	• From the homepage: www.cohere.ca users are able to choose the areas that interest them: administration, online courses or research	• At this time users are unable to customize the portal using a username & password; this will be necessary before the integration goals can be achieved
Transparency		• For the moment COHERE is unable to forge direct links with external communities; however, this is a goal for many Canadian universities and it is inevitable that COHERE will create linkages with community groups interested in online learning
Accountability		• Accountability is a necessary condition in order for the goals of integration & accessibility to be achieved, by putting the power of choice with the learner we are opening ourselves up to public scrutiny of many aspects of university activities
Learning communities / Flattened hierarchies / Collaboration	• COHERE has created a learning community of Canadian universities engaged in online learning; this community is rapidly expanding to include other institutions and individual researchers and learners • COHERE has eliminated the need for multiple levels of bureaucracy by creating a framework in which users can realize a single-entry point for all their learning goals	

REFERENCES

Advisory Committee for Online Learning. (2001, February). *The e-Learning e-Volution in Colleges and Universities*. Retrieved April 22, 2001, from http://www.cmec.ca/postsec/evolution.en.pdf.

American Federation of Teachers (2000, May). *Distance Education: Guidelines for Good Practice* (Report prepared by the Higher Education Department, Program and Policy Council). Retrieved May 1, 2001, from http://www.aft.org/higher_ed/downloadable/distance.pdf

Archer, W., Garrison, R. & Anderson, T. (1999). Adopting disruptive technologies in traditional universities: Continuing education as an incubator for innovation. *Canadian Journal of University Continuing Education*, 25(1), 13-30.

Bakken, C. & Armstrong, S. (2000). *Score: Webpage Evaluation Criteria*. Paper presented at Challenge CTAP Region IV Technology Leadership Academy and the Institute for Research on Learning. Retrieved November 17, 2001, from http://www.ux1.eiu.edu/~cfmgb/web.htm.

Bates, T. (2000). *Managing Technological Change: Strategies for College and University Leaders*. San Francisco: Jossey-Bass Publications.

Batson, T. (2000). *Campus Portals and Faculty Development*. Paper presented at Syllabus 2000: New Dimensions in Educational Technology Conference, Boston, November.

Boettcher, J., & Strauss, H. (2000, January). What is a portal, anyway? *CREN TechTalk Series*. Retrieved November 15, 2001, from http://www.cren.net/know/techtalk/trans/portals_1.html#Gartner.

Campbell, J. (2001). The case for creating a scholars' portal to the Web: A white paper. *Portal: Libraries and the Academy*, 1(1), 15-21.

Carey, T. (2000). *COHERE Memorandum of Understanding*. Waterloo, Ontario: University of Waterloo.

Chickering, A. & Erhmann, S. (1996). Implementing the seven principles: Technology as a lever. *AAHE Bulletin,* (October), 3-6. Retrieved May 5, 2001, from http://www.tltgroup.org/programs/seven.html.

Cooper, A. (2001). *The ZDNet Developer: Church of Usability*. Retrieved November 15, 2001, from http://www.zdnet.com/devhead/stories/articles/0,4413,2780257,00.html.

Copeland, Lee (2001, June). *Ford launches massive corporate portal*. Retrieved November 15, 2001, from http://www.computerworld.com/cwi/story/0,1199,NAV47_STO61399,00.html

Daugherty, M. & Funke, B.L. (1998). University faculty and student perceptions of Web-based instruction. *Journal of Distance Education*, 13(1), 21-39.

Denning, D. & Smith, P. (1998). A case study in the development of an interactive learning environment to teach problem-solving skills. *Journal of Interactive Learning Research*, 9(1), 3-36.

Diaz, D.P. (1998). CD/Web hybrids: Delivering multimedia to the online learner. *Journal of Multimedia and Hypermedia*, 8(1), 89-98.

Eisler, D.L. (2000). Campus portals: Supportive mechanisms for university communication, collaboration, and organizational change. *Syllabus Magazine*, 14(1). Retrieved November 15, 2001, from http://www.syllabus.com/syllabusmagazine/sep00_fea.html.

Erhmann, S. (2000). *Evaluating Campus Portals - Key Ideas*. Syllabus 2000: New Dimensions in Educational Technology Conference, Boston, November.

Everhart, N. (1996). *WebPage Evaluation Worksheet*. Jamaica, NY: St. John's University, Division of Library and Information Science. Retrieved November 15, 2001, from http://www.duke.edu/~de1/evaluate.html.

Ford, R. (2000, June). *Recommendations on Portals*. Missoula, MT: Ad Hoc Committee on Portal Evaluation, University of Montana. Retrieved November 15, 2001, from http://www.cs.umt.edu/AVPIT/portal1.html.

Frontend.com. (2001, May). *Why People Can't Use eLearning: What the eLearning Sector Needs to Learn About Usability*. Retrieved November 29, 2001, from http://www.frontend.com/usability_infocentre/index.html.

Galant, N. (2000). *The Portal for Online Objects in Learning (POOL): An Advanced eLearning Solution*. Paper presented at TeleLearning NCE 5th Annual Conference, Toronto, ON.

Gilbert, S. (2000). *Portal Decisions Demand Collaboration: Can Portals Support it?* The TLT Group. Retrieved November 15, 2001, from http://www.tltgroup.org/gilbert/SyllabusCol2.htm.

Gillham, M. & Buckner, K. (1997). User evaluation of hypermedia encyclopedias. *Journal of Educational Multimedia and Hypermedia*, 6(1), 77-90.

Grassian, E. (1995). *Thinking Critically About World Wide Web Resources*. Retrieved November 17, 2001, from http://www.library.ucla.edu/libraries/college/help/critical/index.htm.

Guzdial, M. & Kehoe, C. (1998). Apprenticeship-based learning environments: A principled approach to providing software-realized scaffolding through hypermedia. *Journal of Interactive Learning Research*, 9(3/4), 289-336.

Ito, M. & Nakakoji, K. (1996). Impact of culture on user interface design. In del Gado, E.M. & Nielsen, J. (Eds.), *International User Interfaces* (pp. 105-126). NY: John Wiley & Sons.

Looney, M. & Lyman, P. (2000). Portals in higher education: What are they, and what is their potential? *EDUCAUSE Review*, 35(4), 28-36.

Marchionini, G. (1992). Interfaces for end-user information seeking. *Journal of the American Society for Information Science*, 43(2), 156-163.

McLellan, H. (1997). Creating virtual communities via the Web. In Khan, B.H. (Ed.), *Web-Based Instruction* (pp. 185-190). Englewood Cliffs, NJ: Educational Technology Publications.

Nelson, L.M. & South, J.B. (1999). *A Process Model for Guiding Naturally Effective Collaborative Problem Solving.* Paper presented at the annual meeting of the Association for Educational Communication Technology, Houston, Texas, February.

Norman, M. (1999). *Portal Technology: Into the Looking Glass.* Paper presented at the 1999 Portal Technology Symposium I, Baltimore.

Oliver, K. (1999). *Computer-Based Tools in Support of Internet-Based Problem Solving.* Paper presented at the annual meeting of the Association for Educational Communication Technology, Houston, Texas, February.

Open and Distance Learning Quality Council. (2000, February). *Standards in Open and Distance Learning.* Retrieved November 15, 2001, from http://www.odlqc.org.uk/.

Paadre, H. & King, S. (2000). *Electronic Community and Portals.* Retrieved November 17, 2001, from http://www.mis2.udel.edu/ja-sig/holycross.doc.

Phipps, R. & Merisotis, J. (2000). *Quality on the Line: Benchmarks for Success in Internet-Based Distance Education.* Washington, DC: Institute for Higher Education Policy.

Ragan, L.C. (1999). Good teaching is good teaching. An emerging set of guiding principles and practices for the design and development of distance education. *Cause/Effect*, 22(1), 20-24.

Reeves, T., & Reeves, P. (1997). Effective dimensions of interactive learning on the World Wide Web. In Khan, B. (Ed.), *Web-Based Instruction* (pp. 59-66). Englewood Cliffs, NJ: Educational Technology Publications.

Usability First. (n.d.). *Introduction to Usability.* Retrieved November 17, 2001, from http://www.usabilityfirst.com/intro/index.txl.

W3C World Wide Web Consortium. (2001, August). *Web Content Accessibility Guidelines 2.0.* Retrieved December 7, 2001, from http://www.w3.org/TR/WAI-WEBCONTENT/.

Western Interstate Commission for Higher Education (WICHE). (1997). *Good Practices in Distance Education.* Boulder, CO: Western Cooperative for Educational Telecommunications.

Websites used as exemplars in this chapter:
COHERE. Retrieved December 7, 2001, from http://www.cohere.ca
Unext. Retrieved December 7, 2001, from http://www.unext.com
Fathom.com. Retrieved December 7, 2001, from http://www.fathom.com/

Chapter XI

Building a Campus Portal— A Strategy that Succeeded

Anne Yandell Bishop
Wake Forest University, USA

ABSTRACT

In 1997 Wake Forest University began the project of building a suite of personalized Web services that is now known as a portal. By July of 1998, we had a fully implemented and successful intranet that delivered almost one hundred personalized, filtered Web services to all students, faculty and staff, plus those alumni and parents who applied for a free account. Being on the leading edge of such an effort meant that we discovered on our own what works and what does not, without influence from portal vendors or benefit of advice from our peers at other institutions. This paper discusses successful design and implementation strategies that may be useful to others who are considering a portal solution.

The term "portal" has become so widely used that its very definition has become a major challenge for institutions that are beginning to discuss whether or not they should have one. Software vendors have developed their own definitions as part of their marketing strategies. Combining observation with experience, I define a portal to be a secure application that provides a single personalized gateway to institutional information and services. The personalization feature sets a portal apart from websites that merely specialize in their focus. Personalization is achieved through requiring authentication and using information from the university's data stores to tailor the content uniquely to each individual.

At Wake Forest University, we embarked in 1997 on the project of building an extensive suite of personalized Web services. By the time portal vendors began arriving on our doorstep in 1998 and 1999, we already had a fully implemented and successful intranet that delivered almost one hundred personalized, filtered Web services to all students, faculty and staff, plus those alumni and parents who applied for free accounts. Rather than research and statistics, this chapter contains a frank discussion of challenges, issues, successes, failures, and what we have learned about implementing and maintaining a portal. Though goals and environmental factors vary across institutions, our experiences can provide valuable lessons for those who are now evaluating vendors' offerings and trying to develop a portal strategy.

BACKGROUND

When the idea for WIN, the Wake Information Network, was conceived in the fall of 1996, there were no definitions to either guide us or limit our vision. Wake Forest was in the first year of its ubiquitous computing plan, having distributed IBM ThinkPads to every entering freshman, more than half of the faculty and many staff members. The plan called for continuing this distribution to each entering class, so that within three years every student, every faculty member and all staff members who used a computer would have a ThinkPad with a full-featured suite of standard software. The plan included replacing student and faculty computers every two years, so the university's ongoing commitment to enabling the use of technology on campus was clear. Though the plan was focused on the use of technology in academic areas, these abundant resources created a fertile environment for the growth of bold ideas.

The idea for the portal was triggered by our university president's mandate to eliminate the long lines at our arena-style class registration. Students were already using their ThinkPads for class work and communications with their instructors and each other. Extending computer use to other facets of student life was a natural next step, though a challenging one. As manager of administrative computing services, I was pondering the complexity of this task when our student records software vendor, Software Research Northwest, Inc. (now part of Sungard Bi-Tech, Inc.), announced a product that provides Web-based class registration. The product, IRISLink, included other Web services such as class rosters, class schedules, grade reports and demographic information with the personalization and filtering capabilities that are now characteristic of portal products. I proposed to implement these services and expand the concept by building additional Web services and infrastructure around IRISLink to improve communication and convenience for the entire university family.

IMPLEMENTATION ISSUES

Getting Started

Continuing discussions on listservs and the popularity of portal development as a topic at higher education conferences seem to indicate that one of the biggest hurdles in building a campus information portal is defining the project and getting it off the ground. The following factors helped us overcome this problem:

1. I submitted a formal proposal for project funding to the vice president for finance and planning. The proposal included a timeline and numerous examples of services the portal might include. The proposal's acceptance represented a major commitment at a high administrative level. Letting the project flounder in endless rounds of discussion simply was not an option.
2. The project was given high priority and was funded and staffed for success. Developing the concept and managing the project became my only job. I hired a software engineer and a part-time graduate student whose only responsibilities were directly related to developing the portal.
3. Our 3-person team had the authority to make the decisions and purchases necessary to deliver the promised services. Though I deliberately chose a collaborative approach to developing the portal's services, I was never hindered or delayed by a requirement to have decisions approved by a committee.
4. Early in the project's organizational phase, I chose and announced a launch date for the portal. We made adjustments as needed to the list of services that would be included in that initial launch, but we never changed the target date. This strategy strengthened the commitment and created the sense of urgency needed to keep the project moving.

Balancing Tradition with Progress

Another challenge in building a campus information portal is managing change by choosing a portal strategy that moves the campus toward its goals while preserving enough tradition to ensure acceptance. An environmental scan to raise our awareness of Wake Forest's defining characteristics and competitive advantages provided the guideline for our strategy. Wake Forest is a relatively small private university, with an enrollment of fewer than 5,500 students and a tradition of highly interactive relationships among faculty and students. The vast majority of our students live on campus; distance education is not part of our mission. Students expect and receive personal attention both in class and in their dealings with staff members who provide administrative services. Wake Forest is a place where strong relationships are forged and the phrase "university family" has meaning. It is also a place where change is often viewed with skepticism, decision making is usually collaborative and exceptions to rules are not uncommon. A one-size-fits-all portal that masked our strong identity and respect for individuality would not have been successful for us even if it had been available for purchase at that time. While IRISLink jump-started our progress by delivering functionality we needed, its architecture allowed us to seamlessly integrate the non-IRISLink services we developed. In addition, it gave us the flexibility to expand in directions of our choosing, to add new services when the timing was right for us, and to define the scope, appearance and navigation methods for a custom fit to our campus. Choosing a build-*and*-buy model allowed us to create our own balance between tradition and progress.

Involving the Right People

It was neither possible nor desirable for my small project team to build a full-featured intranet in a vacuum. We needed assistance, buy-in and decisions from administrators in areas where the institutional data were maintained and where business processes were likely to change when the portal became operational. We also needed help in gathering input from the campus community regarding information and services that we should include. Therefore, a critical early step was to share the vision with these administrators and create a coordinating group that could help drive the project to completion.

The project team had spent its first few weeks creating a test environment using copies of our institutional databases to populate the IRISLink services. This gave us a core set of operational Web functions to spur our imaginations and help us determine what was technically possible. We invited administrators from areas that controlled most of our institutional data to a demonstration of these core services. This visual presentation was instrumental in clarifying for this audience the somewhat nebulous concept of an intranet. Seeing familiar information in a secure Web environment created a lot of excitement among those present, and quickly resulted in a list of volunteers to serve on the coordinating committee. The group chose and began using WIN as the name of the portal in conversations, communications and meetings. This became increasingly important as the circle of those involved widened and name recognition for the project became an asset.

Exposing Data Issues

Populating the test environment with actual data served another valuable purpose by allowing us an early private view of how our institutional data looked in a Web environment. This revealed some data issues immediately, including the following:

- Data entry practices that are appropriate for a paper-driven environment are often inappropriate for a real-time Web environment. For example, recording selected data in all capital letters makes it more noticeable on printed reports but completely unacceptable for Web publication through the portal. Similarly, abbreviations that were used routinely to save space on printed reports were unfamiliar to the broader audience seeing those data through the portal.
- A portal changes the data distribution model radically. Paper reports usually have limited distribution because of paper costs and storage space considerations. Publishing information through the portal makes it both economical and easy to distribute information to a wide audience. Policy must replace mere practicality in making data distribution decisions, and in many cases the policy has never been defined. For example, it had been neither possible nor practical to distribute directories or reports containing student and staff photos. With the portal, though, it became easy to include photos in the internal campus directory. No policy existed to address the privacy issue that would arise if we published these photos in a secured environment.
- Outdated information becomes glaringly obvious. A portal can necessitate changes in long-standing business procedures in offices where data entry is a major activity. Changes such as address updates can no longer wait in a stack on someone's desk until they are needed for a major mailing or a printed report. Campus constituents can view personal information around the clock through the portal, and they expect changes to be made in a timely manner. This forces the staff responsible for those changes to rearrange their workload to make these updates a daily task rather than an occasional one.

Having representation from many business units in the coordinating group was essential to identifying these problems. Each group noticed issues with the data under its control, issues that may have escaped the notice of the project team. The tasks of developing and enlisting support for new policies, procedures and standards were a natural outgrowth of the coordinating group's examination of the data.

Gathering Information from Campus Constituents

With a foundation of sample services in the test account and a vision clarified by many discussions within the coordinating group, we began a series of focus group meetings to gather input from the campus at large. We invited a cross-section of our campus to participate, creating nine groups of approximately a dozen students, faculty or staff members. We conducted each focus group session in the same way:

1. Each meeting began with a demonstration of the sample services, because as yet the public had no concept of an intranet or a portal. The demonstration helped move the groups from blank stares to a dawning understanding of what we were undertaking.
2. We asked each group the same question: what hassles that you routinely encounter as part of your campus life might be addressed through a similar online service?
3. We recorded and discussed all ideas and suggestions within the group in order to gain an understanding of the perceived benefits and problems associated with each.
4. Each group prioritized its item list through weighted voting, where each person could assign point values to five items in order of importance.

The coordinators used the prioritized requests to establish the master list of services whose implementation would be considered by the project team. In some cases, we gathered additional information through personal interviews with deans, vice presidents or other officials who were familiar with laws, policies, Board of Trustees' mandates or traditions that we must not breach.

One common method of gathering information that we did not use was surveys. In 1997 and early 1998, there was still no concept of a portal or intranet in the minds of our campus constituents, so demonstrations and face-to-face exchanges were necessary in order for us to get meaningful feedback.

Selecting Services to Implement

As previously mentioned, the project team rather than a committee had the authority to make final decisions regarding the portal. The team looked at each request on the focus groups' lists and evaluated its technical feasibility before making implementation decisions. Considerations other than popularity become important when requests are thoroughly analyzed, and these factors can determine the success or failure of a service. These considerations and examples are worth examining for anyone who is considering implementing a portal.

The first consideration is whether the data required for the service are available and are already maintained as part of someone's routine. If not, the service is a questionable candidate for successful implementation. For example, one feature that students suggested was a listing or calendar of entertainment and social events in the local area. In order to provide this service, someone would have to gather and update this information on a regular basis. There was no one to assume this responsibility, so we did not consider this service for implementation. As the popularity of the Internet has grown in the years since 1997, a service provider in our city has developed just such a site, and we now offer a link to it from WIN.

We bent this data-availability rule in order to develop one of the most requested services among faculty and staff, and as a result we learned a valuable lesson. The request was for a meeting-room finder, through which individuals could search for campus meeting space with specified characteristics and make a request to reserve the room. The immediate challenge was the lack of central control over meeting spaces on our campus. Responsibility for various rooms is widely distributed and there is no inventory of available meeting spaces. Due to the overwhelming popularity of the request, the coordinating group gathered information on all these spaces and who controlled them, then convened a meeting of these people to discuss the idea of collectively managing the information through WIN. No one was willing to relinquish control of meeting space to someone else, but all were willing to participate in maintaining their own information in a common database. The project team developed a service that allowed each person who had space management responsibilities to enter and update information on his or her designated spaces and to remove rooms from the inventory. This worked well during the first year after WIN's implementation, but people soon forgot about maintaining this information because it never became part of their daily routine. Staff responsibilities for various spaces changed, the inventory itself changed as existing buildings were remodeled and new buildings were built, and the database gradually became more out of date and less useful. The service was finally dropped from WIN after three years.

Two similar services that were successful illustrate the subtle differences between what works and what does not. These examples are the WIN Announcements service, which allows selected administrative offices to post announcements for entire constituent groups (faculty, students, alumni, staff or any combination of these), and the Forms and Documents Library, which allows departments to post downloadable forms for use by members of the campus community. Both services involve new data management tasks required of a widely diverse group across many departments. WIN Announcements are sometimes a supplement to existing communication methods such as sending memos, but are more often a replacement for ineffective methods such as posting signs around campus and word-of-mouth. The Forms and Documents Library allows a department to ensure that its most recent version of a required form, such as a travel reimbursements form, is always available for campus constituents. Unlike paper forms, the electronic forms can contain error-reducing features such as totaling or calculating travel allowances based on mileage. Both WIN Announcements and the Forms and Documents Library impose new tasks on those staff members in each department who manage posting the announcements or forms, but the departments reap benefits in terms of enhanced communication and greater accuracy. These benefits rewarded staff for taking on the new service management tasks and making them part of their routines; with the meeting space management system, there had been little or no payback for the staff to take on new data entry tasks.

A customer service orientation is a second consideration in deciding which services to implement. In order to be considered for WIN, a service had to provide a visible direct benefit to large numbers of the university family. Cost savings and efficiency gains in business offices might result from some services, but might not be sufficient reason for their development.

Some services result in marginal efficiency gains at best for the business staff, but huge benefits for its campus customers. For example, getting a permit for parking on campus had always required that faculty and staff members fill out a multi-page bubble sheet that was available only in the Parking Office, located in an out-of-the-way building on the edge of campus. Though the staff members in that office were perfectly satisfied with their scanning equipment and their process worked well for them, it was a time-consuming inconvenience for faculty and staff members who had to go out of their ways to register a vehicle or make any change in their registration. Handling this process online was high on the list of requests from focus groups. We implemented the WIN Vehicle Registration service, enabling faculty and staff members to submit these applications as Web forms and receive their parking stickers in campus mail a few days later. The cost savings were minimal for the Parking Office staff because they already had an automated process, but the benefits for the campus community in time and convenience were great. This orientation toward customer service helps ensure that WIN provides value for the majority of our campus family members, and guides us in setting priorities for the development team.

A third and very important consideration in deciding which services to implement is whether or not the business offices whose processes are affected by the service are willing to participate and able to reengineer their procedures as needed. The parking permit again serves as an example. Permits for faculty and staff are free, so the process change for implementing the faculty permit application as a Web service did not impact Financial and Accounting Services (FAS). However, permits for students involve a complex payment scheme and thus also involve FAS. During WIN's initial development period, changes were underway in the FAS office, both in their finance software and the methods and procedures they used for accepting credit card payments. It simply was not a good time for them to work with us on plugging a student permit service into WIN, and so that service was not part of the initial implementation.

A final consideration in choosing services is the level of effort required to implement the service. This is a classic case of "picking the low-hanging fruit." We delivered many IRISLink services regardless of their priority because there was minimal time and effort involved in doing so. We also delivered some lower-priority services such as a birthday calendar whose development we could assign to a work-study student with minimal programming skills. We were later surprised by the success and popularity of some of these services. We kept the number of complex, time-consuming services to a minimum because of our timeline for the initial launch of the portal. The end result was that we delivered a well-rounded, full-featured portal with something of interest to almost everyone.

Technical Design Issues

Technical design issues fall into two categories: design of individual services and overall technical design principles. To manage the design of individual services, we invited focus group members and their colleagues to meet with the project team to provide more detailed requirements and critique the results. This approach was particularly successful with faculty and academic advisers, who met with the project team weekly during the development of instructor and adviser services. Their direct input resulted in services with exactly the information faculty and advisers need arranged in the ways they want it, ways the project team would not have anticipated.

The second category of technical design issues encompasses several principles that contribute to making the portal self-sustaining and reducing maintenance for the technical team. These principles include minimizing access control tasks, building tools to put management tasks into the hands of others, making the portal self-cleansing, minimizing the effort involved in changing a service's presentation and facilitating the addition of new services. Because these issues are so fundamental to the success of maintaining the portal as a viable, long-term tool, examples are given below to clarify each concept.

Managing access control for various services can become a full-time job without planning and design. One strategy we used for minimizing access control tasks for the technical team was to successfully incorporate the task into someone else's routine. As an example, there were a number of WIN services for which we needed to grant access to either the chair or the administrative assistant in each academic department. No one maintained database markers to indicate the identity of members of either group. Thus there was no information available to WIN for access-control decisions. However, the registrar's staff kept such information on lists that they used frequently. The WIN team developed a method by which the registrar's staff could keep this information in database markers instead of lists. Now many WIN capabilities are granted automatically from those markers. The benefit for the registrar's staff is two-fold: they can communicate electronically with these groups through both WIN and their student information system, and the department chairs and their assistants can use WIN to get information for themselves that they formerly would have had to request from the registrar's staff. Thus we see again that the key to success in getting people to take on new access management tasks for the portal is to make sure the portal provides rewards in return.

Some services, particularly data-management tasks such as processing address changes, must be granted to one or more people across several departments. In these cases, there must be an access control list to determine who in each department has those specialized WIN privileges. We sought a strategy that would minimize the task of maintaining access control lists for the Information Systems staff. To accomplish this, we developed a Group Permissions WIN service for managing access rights to other services. It allows the Information Systems staff to quickly grant control of a specified service to a key contact in a department; that contact then gains the capability to grant access rights to others within his or her department to the specified service. This reduces the maintenance burden for Information Systems to one departmental contact, and puts further control into the hands of trusted contacts who know which of their colleagues should have the WIN capabilities in question. The lesson in this example is to develop a tool for making tasks easy to distribute to others when the tasks cannot be avoided entirely.

Another essential factor in making a portal self-sustaining is to make it self-cleansing by ensuring that old information is deleted on a regular basis. WIN has a number of features that allow an individual to post information, such as a Ride Board for students, Classified Ads for everyone and a Used Textbook Exchange for students. A critical design feature of each of these services, as well as the WIN Announcements service, is that entries expire and are automatically deleted from the system by a WIN process after a specified period of time. This relieves both the technical staff and the individual posting the information from the task of remembering to delete postings, and keeps WIN's information fresh.

Minimizing the work required to change a screen's text or appearance is also a key factor in reducing maintenance for the technical staff. To facilitate such changes, the project team separates the business logic from the scripts that control Web presentation. Developers can make changes to a screen's text or appearance quickly without modifying program code, and can grant designated staff members in some departmental offices the capability to manage the appearance of their own services. For example, one service allows alumni to apply for free WIN accounts and Wake Forest e-mail forwarding addresses for life. Designated staff members in the Alumni Office can manage the wording and content of the application page using the Web presentation scripts. Designing services and developing tools to put control in the hands of those who have the information readily at hand minimizes ongoing maintenance for the portal developers.

The method for adding new services is another design consideration in developing a low-maintenance portal. WIN's dynamic menu generation feature allows the project team to add new services with minimal effort, making the portal highly extensible. Any part of the WIN menu can automatically expand to many items or disappear entirely depending on what the user is entitled to see. However, no one has the labor-intensive task of maintaining HTML links as we add and remove WIN services for various groups. We can plug in a newly developed service within minutes by simply adding entries to tables containing names of services and groups entitled to receive those services. This was a design feature of IRISLink that we adapted and extended to our own development.

Tools and Technology Decisions

The pace of technological innovation coupled with the many options available when choosing a server platform, development languages, Web server software and security tools can easily freeze a project team into inaction. Our development team simply made initial choices based on their skill sets and their research into security, scalability and extensibility. However, no choice was viewed as permanent. The team continued to monitor the latest developments during the initial implementation period and made adjustments to those initial choices when needed. Over time we have successfully moved parts of WIN from NT servers to Unix servers based on the skills and expertise of the project team members. We have adopted new technologies such as XML as they have emerged. We have successfully migrated some early services to a new database engine in keeping with a change in other campus systems. The strategy for anyone who is considering developing a portal or purchasing a portal product should be to choose tools, hardware, software and products that your staff can support, then remain open to change as circumstances dictate.

Managing Risk

The risk factors for a portal service are similar to risk factors for traditional business applications. Risk increases with the complexity of the service, the number of departmental boundaries crossed, the number of interrelated processes involved, the level of opposition to change among those who are directly involved and lack of high-level support for the change. These factors can be dealt with in traditional ways. An important difference, however, is the high visibility of Web services, which adds an even higher risk factor to portal development. Unlike the situation with traditional business services that are typically viewed or used by a few people, the audience for Web services can be hundreds or thousands of people. We took several steps to manage this high-visibility risk. First, we counterbalanced the risk inherent in the data itself with the complexity of the service. For high-risk data such as students' grades, we implemented simple view-only services for the initial launch of WIN. More complex interactive services that involved submitting new data or changing existing data were reserved for lower-risk information, where a malfunction would be of much less consequence.

Our second strategy for managing risk was to take advantage of our vendor's expertise and resources for high-visibility, complex applications such as class registration. Through detailed planning with the registrar's office and others directly affected by Web-enabling class registration, we were able to adapt the software and our procedures in such a way that the essential functionality of the vendor's product was unchanged. After thorough testing on our own, we conducted our first registration as a pilot with 100 students rather than the entire student body. The pilot uncovered some unanticipated data problems, but recovery was not difficult because of the small number of students involved. By the time of the first full registration for all students, we had solved those problems and gained several more months of experience in monitoring and fine-tuning complex applications. Since the initial launch of WIN, we have developed additional high-risk services such as student payroll timecard submission and approval. As our experience increases, the risk involved in deploying such services decreases.

Protecting an individual's privacy while granting access to information is of utmost importance on a college campus. The risk associated with violations of the Family Educational Rights and Privacy Act (FERPA) is very high. To address this risk, the team designed WIN's directory services to follow existing privacy settings managed by the various university records offices. For information not already covered by those privacy settings, the team developed a WIN Preferences service that enables each individual to manage his or her WIN password and to specify whether or not to display certain information in WIN services.

Another risk-management strategy that we employed addresses the well-known risk associated with unpredictable usage levels for Web services. Nothing generates unfavorable publicity as quickly as a Web server that cannot respond to demand. We had no history on which to base predictions for usage, so we deliberately launched WIN with little fanfare during the summer when the majority of our students were gone. This gave us time to monitor performance and usage patterns and learn to manage them while our population was low and unanticipated downtime could be tolerated.

The final risk-management strategy was ensuring that there were alternatives to WIN services for some period of time after its initial implementation. For example, faculty members could still request and receive paper copies of their class rosters; they could submit their grades on scan sheets until they became comfortable with submitting them through WIN; academic advisers could still get paper copies of students' degree audit reports. In addition to relieving pressure on the technical staff, this approach allowed acceptance of WIN by the campus community and confidence in its capabilities to grow before it became the *only* method for accomplishing any task.

The Ongoing Challenge—Growing Demand

With success comes increasing demand. We launched WIN on a largely unsuspecting campus that had never envisioned such a thing and had not requested it. However, students enthusiastically embraced the internal campus directory, with photos included, and discovered that they could find their grades on WIN well before the grade report arrived in the mail. Faculty members found that trips to the registrar's office to get information were far less frequent. Students and faculty members studying and working abroad found it easy to register for the next semester's classes or to submit grades through WIN from anywhere in the world that Internet connectivity was available. Ideas and requests for more WIN services began to stream in soon after its initial implementation. A significant milestone occurred when the university's controller *required* that department heads submit budget requests through WIN rather than on paper. Subsequently, the Romance and Classical Languages faculties requested and collaborated with the WIN team in developing language placement test services for incoming students. True to our design philosophy of making WIN self-sustaining, the team developed a test wizard rather than a test requiring continuing maintenance by technical staff. Faculty members in each language area use the wizard to manage the test's content and scoring rubrics. Faculty, advisers and students now benefit from immediate test results, which are fed into the class registration system. Tests for these languages are no longer given on paper, and analysis is underway for incorporating placement tests for other academic disciplines into WIN. All of these factors are indications of WIN's acceptance as a vital method of conducting business on our campus.

The growing demand has contributed to changes in the organization and allocation of resources in the Information Systems department as the university has changed the way it does business. In late 2001, the WIN team has grown to include several developers plus server and database management specialists who were formerly part of other teams. Server scripts automatically notify staff members when any WIN overnight process fails or when other problems occur that affect WIN's availability. The department has also changed the way it manages requests for new WIN services. To get the project off the ground, we initially used an informal approach. Due to competing demand for resources, however, we now use a more formal request and approval process, though the original criteria for choosing services to implement continue to apply.

SOLUTIONS AND RECOMMENDATIONS

The value of a portal is greatest when the institution determines its own strategy rather than adopting someone else's vision and definition. Though there are many examples of what has worked for us in previous sections of this chapter, the ultimate strategy for your campus must be your own. Based on Wake Forest's experience, I offer some general recommendations for developing a strategy.

First, assess your technical environment. Align your strategy with the existing environment and the institution's strategic plan for computing resources. Just as many of our decisions are driven by the fact that every student and faculty member has a computer with a standard software load readily at hand, your portal strategy must reflect the resources available to your campus constituents. Evaluate vendors' products for fit with your technical environment to be sure you are purchasing services that you can reasonably implement and make available to your campus constituents.

Second, strike the appropriate balance between tradition and change at your institution. Involve people in the project who know and respect the institution's history but are not afraid of change. Although WIN's acceptance is no longer in question, we remain very much aware of our campus culture of collaboration and the need to respect our traditions while quietly enabling change. Evolution rather than revolution is a good strategy for our campus, but may not be right for everyone.

Third, base decisions on what your campus wants and needs rather than on someone else's definition of what a portal should be. The needs of a small residential campus such as ours differ greatly from those of a university with a large percentage of non-resident or adult students. For example, distance education is not a part of our mission; it may be vital to yours. This may dictate the need for a strategy that is quite different from ours. That decision must be the result of a careful needs analysis.

Base decisions on needs, but also on what your campus can support. Use tools that your technical staff can manage successfully. Take advantage of products and services from vendors with whom you already have relationships and technical compatibility. Implement self-sustaining services based on your campus data stores, using data that someone already maintains routinely. Design services to put control of content into the hands of those who have the knowledge and the motivation to keep it up to date.

Finally, keep two words foremost in your decision-making process: add value. Use your resources to implement services that will add the most value for large numbers of users, services that they cannot get elsewhere.

THE FUTURE

As of Fall 2001, WIN has become such an integral part of the campus fabric that it is now considered an essential service rather than a convenience. Automated processes create accounts as needed daily for any newly admitted student or newly hired employee. More than 10,000 alumni in several countries have WIN accounts, and the number continues to grow. We are considering strategies for making WIN services more easily accessible on hand-held computing devices. The long list of requests for new services ensures that WIN's role will expand in the future.

Beyond our own institution, the trend toward portal implementations appears to be strong. The strategy of community building has been widely adopted by commercial enterprises as well as educational institutions. Commercial enterprises build portals to gain customer loyalty and thus repeat business; the portal enables them to do targeted marketing as well as to retain some personalization in the absence of face-to-face contact. Universities lack the direct profit motive for building a portal, but targeted communication, community building and retaining personalization are important for educational institutions as well as businesses. Technology can easily drive an organization from low-tech, high-touch interactions to high-tech, low-touch interactions with its constituents. A portal's value is in allowing the university to maintain both high-tech and high-touch interactions with its family, who are always just a mouse-click away.

CONCLUSION

A portal is an important asset in an institution's ability to provide services and to meet the growing expectations of students, faculty, staff and alumni. Web-based services are becoming the norm rather than the exception. An institution's ability to compete for the best students and the best faculty is based not only on its academic reputation, but also on its ability to provide a good environment in which to teach, learn and work. Former generations of students and faculty were most comfortable using pen and paper; future generations will be most comfortable using electronic communications and services. A portal's value as a tool for communication and for conducting the business of the college or university will grow as our society becomes increasingly empowered by technology.

SECTION III

VENDORS' PERSPECTIVES

Chapter XII

Portals Unlock the Knowledge that Drives Business Value

Robert Duffner
BEA Systems, Inc., USA

ABSTRACT

In this chapter, Robert Duffner, director of product marketing at BEA Systems, defines the landscape and technologies for tenterprise portals. Beginning with the business drivers that led to the historical creation of the product category, the article goes on to describe the three generations of portal product evolution form simple packaged applications to robust portal platforms. Duffner describes a five-tier framework that defines the technical underpinnings of the ideal portal platform and provides a vision for how the technology will evolve over time.

For the past 30 years, enterprise computing has pursued the goal of information systems that can respond to the ever-increasing pace of business change and the concomitant need for increased information and access. From the requirements planning (MRP) systems of the 1970s, to the enterprise resource planning (ERP) systems of the '80s, to customer relationship management (CRM) in the '90s, each new technology was hailed as the answer to enterprise information access. While these systems offered new functionality and promised higher levels of access and visibility, their value was effectively limited by the need to integrate the new technology with legacy systems and infrastructure.

Throughout the 1990s, new technologies emerged to meet this need for integration: data warehousing, client-server computing, enterprise application integration, information search and retrieval. All these offered the promise of "unlocking" the potential of enterprise information resources and business intelligence. These technologies, in their turn, presented their own limitations, most notably their reliance on proprietary methods of integration. Not only was the cost of maintaining this custom-built integration high, these tenuously linked islands of automation lacked the flexibility to meet the emerging challenges of conducting business over the Web.

At the end of the 1990s, enterprise portals made their debut, combining many of the features of earlier integration technologies. Portals presented new opportunities to extend existing applications as well as a mechanism to quickly deploy new Web-based applications. The first-wave portals were primarily packaged applications, each with a specific focus—for example, benefits enrollment—and a narrow view. These portals typically delivered application services to the organization's different constituencies: employees could look up benefits information, students had access to course and registration information, suppliers could view open requests for proposals.

As portals proliferated within organizations, extending from department to department, managing these development and implementation projects became increasingly complex. While the portal provided benefits to single departments—such as human resources, sales and engineering—the complexity of these solutions snowballed by gradually adding layers of software and independent integration points to be managed by the central IT department. These separate portals evolved from access points into an application deployment framework in and of themselves. They became prey to many of the same integration problems as the earlier applications they were intended to solve.

Today, organizations in businesses of every type are rushing into the development and deployment of portals—a veritable California Gold Rush. The benefits of portal software are widely understood: portals make an organization more accessible to its customers and partners, and offer a way to create identity and competitive advantage through personalized interaction. Industry analysts have often cited the following as benefits that flow from effective portal implementations:

- Reduced costs and investments
- Accelerated business processes and elimination of steps
- Increased sales
- Higher competitive barriers
- Reduced training costs

Portals offer these benefits precisely because of their ability to bring together in a single location a wide variety of information sources—both internal and external—and applications. As a result, today's enterprise portals are unifying platforms that organizations can use to leverage their existing technology investments. To realize this potential, a new generation of portal software is on the horizon. This next generation solution is built on reusable components, thus transforming the role of the portal from a single-purpose application into a delivery platform for multiple and varied sets of application services: a *portal platform*. This chapter will discuss and define the portal platform technology elements required to deploy next-generation, enterprise portals.

BACKGROUND

The corporate portal market is one of the most rapidly growing segments in the computer industry today—according to Forrester Research the median cost of planned portals is as high as $1 million (Gillett, 2001). Portal software sales were $252.1 million in 2000, and IDC (formerly International Data Corporation) sees the number rising to $2.4 billion by 2005 (McDonough & Morris, 2001). IDC also projected that U.S. business adoption of portals would jump from 12.7% in 2000 to an estimated 20.1% in 2001.

Because the portal market is relatively immature, there is considerable variability in ideas about what constitutes a portal. However, the essential characteristic of a portal is that it provides an access point to applications and services directed at a specific audience and designed for specific business processes. Without a defined audience, a portal has little meaning. So, one of the essential characteristics of portals is that they are inherently personalized to address a certain audience—be that employee, supplier, partner or customer. In fact, many organizations maintain portals for segments of employees, suppliers or customers. As portals become more pervasive, it is clearly desirable for any given enterprise to deploy and manage all portals on a single platform. To achieve this, however, the underlying infrastructure software must be able to adapt to the varying requirements of different types of portals.

Types of Portals

Portals can be categorized based on the intended user: internal audiences such as employees, students or faculty, and external audiences such as customers, alumni, business suppliers and partners. The type of portal dictates the content, integration requirements and the functional capabilities that are needed. Several examples follow.

Employee Portal: Employee-focused portals provide organizational operations information that needs to be shared with staff. These are typically first-generation portals (see *Evolution of Portals*, below), although today more employees need access to business and organizational data to make informed decisions. As a result, employee portals are under increasing pressure to expose organizational processes, systems and data to certain employees. A classic example of an internally focused portal is a corporate intranet site containing benefits information that employees can examine at their convenience. Other examples of employee portal applications include:
- Report management and distribution
- Market intelligence
- Function-specific kiosk; for example, benefits enrollment
- Cross-repository information search; for example, linking product specifications and marketing copy
- Problem tracking
- Help desk

Good employee portal applications can provide significant benefits to an organization. First, they decrease the IT backlog by eliminating many requests for special reports. Second, they improve access to business-critical information, thereby enabling more prompt and effective action. Third, employee portals also facilitate cross-departmental collaboration to increase overall organizational efficiency. And finally, they improve the quality and timeliness of communications across the organization.

Consumer Portal: Consumer-oriented portals provide corporate information that needs to be shared with customers. Thus, they require additional integration and performance capabilities because customers want 24/7 self service as well as access to internal data like product inventories, specifications and shipment status. A classic example of a consumer portal application is a customer order application (also known as a customer self-service application) that allows a consumer to purchase a company's products over the Internet, usually without assistance. Other examples of consumer applications include:
- E-commerce services
- Portals offering customers personalized services, usage-based recommendations, etc.
- Warranty registration
- Support services

Customer self-service applications can reduce the cost of doing business through automation, while increasing opportunities for organizations to interact with visitors through online tools, applications and communities. These applications can deliver marketing messages, reinforce corporate branding and ultimately differeniate the customer experience from that of other businesses. The result is increased customer satisfaction and increased opportunities to sell more products, services and content.

Business Portal: Business partner- and supplier-focused portals provide corporate information that needs to be shared with other businesses. In the increasingly collaborative business environment, business portal applications are growing popular and demand for business-focused portals has helped create a need for next-generation portals. Some examples of business-to-business applications include the following:
- Inventory visibility
- Channel management
- Private trading networks
- Shared application services
- Order management
- Customer (business) self-service

Because customers in this environment are other businesses, the benefits that apply to customer self-service applications likewise apply to business applications. In addition, by fostering a more collaborative relationship with business partners, business-oriented portals can promote significant improvements in responsiveness by synchronizing supply and demand chains.

EVOLUTION OF PORTAL PRODUCTS

When organizations choose a platform for portal development and deployment, that platform should inherently support evolving requirements. Gartner, an industry analysis and consulting firm, describes this evolution in terms of first-, second- or third-generation products; applying this construct to portals is helpful because it allows us to envision advances by evolutionary steps from limited-function portal software to a comprehensive portal platform (Phifer, 2001).

Generation One

First-generation portal products were focused on content. Characteristics of these systems include:
- Search and categorization
- Content management
- Customized content

- Limited application development and integration
- Limited infrastructure

Generation One products are characterized by many "out of the box" features that facilitate rapid deployment. As these Generation One products evolve, the major challenge is to retain that ease of deployment, while adding the robust capabilities necessary for Generation Two. This is not easy because these packaged applications were never intended to supply more than a specified set of Generation One features. They typically do not supply the application infrastructure and tools to assemble software flexible enough to meet dynamic needs. Further, as portals become an increasingly important part of an organization's "brand," packaged applications do little to help create competitive advantage because all competitors can have the same applications.

Generation Two

Second-generation portal products can be viewed as portal versions of e-business platforms, and have the following characteristics:
- Robust application development and run-time platforms
- Robust integration capabilities
- Multi-channel (or wireless) capabilities
- Collaboration capabilities
- Improved first-generation portal features

Vendors that integrate traditional platform and enterprise applications have an advantage in the transition to Generation Two because their core competencies are major features of Generation Two portal products.

Generation Three

Third-generation portal product functionality begins to fulfill the promise of enterprise portal applications by supplying a unifying platform that transforms portals from simple "stove-pipe" integration of individual business areas to the full integration of content and applications across entire enterprises. The most important feature of Generation Three portals is enhanced personalization that tailors the material presented to both the function and the user. In addition, Generation Three portals are characterized by:
- Business process management
- Delegated administration
- Entitlements
- Full e-business application development platform that enables enterprise-level deployment

PORTAL VENDORS

As the portal market continues to grow, three types of portal vendors have emerged: pure-play portal vendors, application integration vendors and e-business platform vendors.

- *Pure-play portal vendors* are focused primarily on selling a packaged set of portal functions, applications and tools, often to a specific market, that allow users to configure portals. Vendors such as Plumtree and Epicentric are examples of pure-play vendors.
- *Application vendors* provide vertical solutions based on an application. These portal products are tied to the underlying packaged applications of the vendor, such as enterprise resource planning (ERP) systems. Examples of these vendors include SAP, Vitria and PeopleSoft.
- *Enterprise platform vendors* provide a single software platform on which to build multiple types of portals. These platforms can provide both the infrastructure software and application framework that enable rapid development and deployment of many types of portal applications. They can also include pre-built components for integration and interaction. Two examples of these vendors are BEA Systems and IBM.

Today, traditional portal vendors are being challenged by suppliers offering enterprise platforms and application infrastructures that can be used across an enterprise IT environment. Typically, the core business of the software vendor defines the functionality and development of the portal software: document publishing, collaboration, higher education or line-of-business applications (ERP, customer relationship management). As enterprise portals become a convergence point for enterprise applications, business processes and e-business, the portal framework becomes the primary delivery platform for all portal services across the enterprise. This implies requirements that extend far beyond application-specific needs. For optimum success of their portal strategies, organizations must ensure that new applications and new information sources can be integrated into the portal as soon as they are deployed.

THE VALUE OF GENERATION THREE PORTALS

While the Generation One portals were delivered as ready-to-run applications with minimal customization, today's environment is a rapidly evolving one in which the portal platform must support the overlapping lifecycles of new and changing applications. First-generation portals were designed for accessing static information. However, visitors to today's portals expect real-time access to applications and back-end systems around the world and around the clock. Customers want self-service; for example, real-time access to inventory and shipping information, or 24/7 access to course work or class enrollment. Today's portals must support and distribute application services into different environments that make up the extended enterprise.

According to a study by Forrester Research (Gillett, 2001), 41% of the commercial sector portal managers they interviewed expected to integrate seven or more systems into their enterprise portal; 22% expected to integrate 20 or more. Forrester further reports that 92% of the organizations they surveyed planned to incorporate existing intranet sites into their portals; 59% planned to incorporate existing application servers. To serve this growing audience, IT professionals in business and in higher education are looking for a single environment, i.e., the more the user can do without leaving the portal environment, the more value the portal delivers.

The need for expanded access has obviated the use of early portal products. Some institutions are looking at portals as the new desktop, or as the glue to their Knowledge Management strategies. The portal becomes a framework for unifying access to multiple types of content while, at the same time, personalizing the interface to specific organizational roles, individuals and contexts (Gillett, 2001). The narrowly focused, application-specific portal model cannot accommodate this. Seen in this light, a portal represents more than a simple access point. It supplies a pervasive infrastructure for realizing increased value from an organization's information systems investments. By providing a single, unified user interface, a single sign-on and personalization at multiple levels, third-generation portals are at the heart of this pervasive infrastructure.

A MODEL FOR A PORTAL PLATFORM
Issues: Meeting Enterprise Performance Requirements

Evolving from Generation One and Two portals to Generation Three is not simply a matter of adding new functionality. To deliver the performance, reliability and flexibility required by a unifying, enterprise-wide platform, Generation Three portals must be architected to deliver these capabilities.

Many Generation One portal products run on Web servers and typically run as Active Server Pages, Java Server Pages or servlets. For this reason, portals that use the Web server environment to run portal components and portal applications will suffer scalability and reliability issues under heavy loads. Generation Two products solve this problem by using an application server, but simply running the portal component on an application server is insufficient. The portal product must leverage the load balancing, failover, monitoring, prioritization and management features of application servers to support high-end, enterprise-class portal operations. Using application servers in this mode is very complex; however, this is required for the highest levels of enterprise-class operations.

If an enterprise portal is to supply a pervasive infrastructure, it must:
- Be delivered on scalable, high-availability platforms
- Support wide-scale access
- Supply intelligent management tools and advanced security across a wide variety of applications

These sophisticated portals demand a portal server infrastructure that can address front-end requirements such as aggregation and presentation as well as back-end requirements such as data updates and security.

Web application servers have evolved to become the de facto foundation of this multi-tier architectural approach. Traditionally, application servers have provided infrastructure, management capabilities and tools. Providing the building blocks of the architecture, the application server supplies the tools to assemble new applications and an infrastructure that offers scalability, extensibility and reliability. Evolving reinforcement of standards such as Java 2 Enterprise Edition (J2EE) provides a stable base of knowledge as well as an existing catalog of components that can be leveraged to build new applications.

Many functions can reside on the application server. First is the business logic execution that includes software applications and business rules. The application server also manages data and information access and integration. Application integration also resides here in addition to critical features that are not addressed by standards--for example, support for distributed, component-based solutions. The application maintains security; all applications access the central repository and use a common security model. Transaction processing capabilities, scalability, reliability, run-time load balancing, the developer interface and data persistence are also essential parts of the application server. In addition, the application server manages distribution and replication of functionality for load balancing, failover handling and the addition of incremental resources when required. A variety of industry vendors and analysts have structured these functions into a portal platform model consisting of interacting layers of application functionality that reside on top of existing systems and infrastructure.

Solution: A Five-Tier Portal Platform Architecture

The Delphi Group (2001) proposes a five-tier architecture:
- The **systems tier** provides the fundamental infrastructure: databases, packaged applications, directory systems, content repositories and other supporting applications.

- The **server tier** runs, manages and maintains Web-based applications, supplying the infrastructure to build and support the applications deployed on the portal platform.
- The **integration tier** enables connectivity between components with rules and logic that govern business processes and application services.
- The **interaction tier** manages connections between the user interface and services and content, supplying the rules and logic governing the behavior of user-to-application connections.
- Finally, the **presentation tier** supports access to applications and content through multiple devices and application forms.

Technologies like J2EE and Microsoft's .Net architecture supply the necessary adaptability to support this comprehensive portal platform. Standards like J2EE Connector Architecture (J2EE CA) and Simple Object Access Protocol (SOAP) offer standards-based connectivity between diverse applications and content, replacing proprietary connectivity and its concomitant overhead. Each of

Delphi Group's 5-Tier Portal Platform Architecture

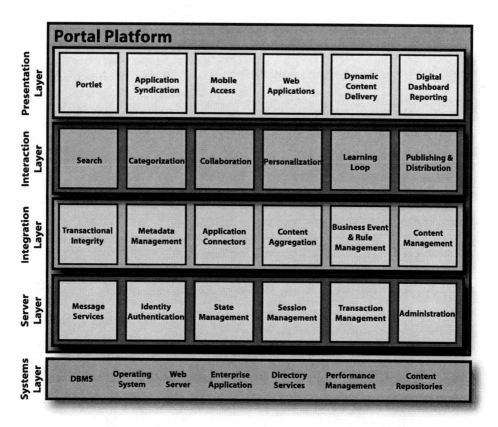

Functional Components Used for Building Applications Within the Portal Architecture

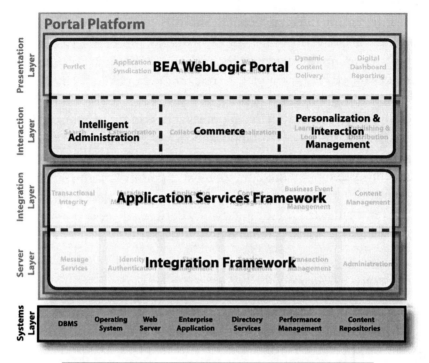

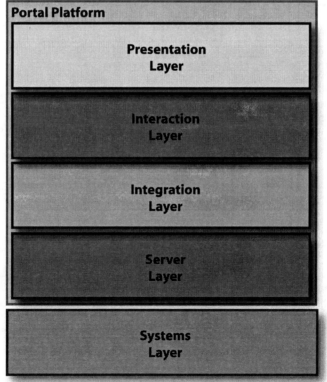

the five layers in turn contains functional building block objects for assembling complete applications. The ability to recombine and redeploy these components fosters the agility that organizations need to build and launch Generation Three portals.

The Portal Platform Software

To build and deploy Generation Three portals on the proposed architecture, five primary software components are needed: portlets, portal foundation services, personalization and interaction management, intelligent administration and integration services.

Portlets

The portal user interface uses a software module called a portlet—also called a gadget or widget—as its primary building block for presentation and content aggregation. These are compact windows arranged on portal pages to provide access to content and applications. The portlet model encourages modular development by allowing new applications to be exposed as portlets and plugged into the portal. Ideally, the portal platform should include some pre-built portlets, providing content or application-specific functionality that minimizes portal time-to-market and maximizes development efforts.

Portlets can be used for any type of content to be included in the portal, such as:
- Research tools
- Collaboration environments
- Content management tools
- Reporting and analysis tools
- Packaged applications
- E-mail
- Site and Web search tools
- Event tracking
- Syndicated content

Standard HTML, Java Server Page (JSP) and Java development tools should be supported for the development of new portlets, and the portal platform should provide an open environment for defining portlet navigation and process flow.

Foundation Services

Foundation Services provide the tools used by developers to create, customize and manage portals, and are the foundation upon which the ultimate adaptability of the portal platform rests. Foundation services should supply the basic set of portal services for user interface and presentation, integration with content management systems, security and commerce services, portal-wide search, application deployment, and scalability and performance. This functionality simplifies complex portal development, maintenance and security, and maximizes the overall effectiveness of IT resources. Following is a discussion of several of the technology components of foundation services.

Portal presentation services provide an easy way of adding application and content functionality through JSP-based portlets. Portals created with these services provide a rich Web-based user interface that can include multiple portal pages. Customization capabilities should allow portlet selection, location and "look-and-feel" to be specified at the end-user, user group and portal levels. This architecture also supplies a foundation for extending portal functionality with custom-designed portlets and portlets provided by partners as well as through new development.

Layouts and skins are the fundamental elements that allow portal page designers to specify the look and feel of the portal and should accommodate any design style. Layouts are "wireframes" that use placeholders to define how portlets will be arranged on the portal page. Skins define the overall look-and-feel of the page, specifying the fonts, colors and icons used by portlets.

Content integration interfaces are a set of interfaces to third-party content management systems. In addition to retrieving content and documents, these interfaces can also be used to personalize applications.

Other elements of foundation services include commerce services such as catalogs, search/browse capabilities, shopping carts and order management, and support for back-end system integration. An integrated search capability should be available for metadata searches of catalogs and content repositories.

Personalization and Interaction Management

Portal presentation services, such as personalization and interaction management, are used by designers to create portal pages using portlets, and are an essential element of a portal infrastructure that can be adapted to both individual users and groups of users. For example, this layer of technology can be extended to marketing and CRM applications to provide a more comprehensive view of the customer.

Personalization and interaction management controls the user experience, defining, customizing and measuring user interaction. These services improve the user experience by providing both implicit and explicit personalization as well as a framework that can be used to better target content and services to the user. Rules-based personalization should support both implicit and explicit personalization using online browsing behavior and explicit profile information to classify visitors and serve relevant content based on profiles, behavior and preferences. Customer segments can be used to classify users by similar attributes to create segments that can be used as the basis for presenting specialized content. Real-time evaluation uses the visitor's current browsing behavior in combination with other sources of information to dynamically supply the visitor with relevant personalization and interaction.

Event and behavior tracking supplies another important component of interaction management by recording page impressions, click-throughs, commerce events such as "add-to" and "removal" from shopping carts, and purchase and order histories. Event data can be used for personalization and publishing online marketing or education campaigns to specific users, and should be accessible using the organization's existing business intelligence tools. Likewise, the portal platform should supply integration with existing online and offline campaign management tools.

Intelligent Administration

Intelligent administration and configuration of the portal platform is critical if it is to fulfill its promise as an enterprise-wide infrastructure.

As the portal platform becomes the organization's primary information system infrastructure, decentralized administration becomes a requirement because it enables user groups to directly fulfill their needs to provide information transactions to their clients while simultaneously reducing administrative backlog. This is helpful for frequently occurring tasks like user management administration. Delegated administration allows administrators to entrust portal administration tasks or subsets of tasks to others, with administrators able to limit the scope of delegated administration to specific portals or specific user groups. Management of the layout, fonts and colors—portal look-and-feel—might be delegated to a design group, for example. Delegation of entitlements allows users with organizational responsibility to dynamically enable portal functionality through access to portlets and portal pages.

Rules-based entitlements should be used to control user access to portal pages and portlets with business rules rather than traditional access control lists based on fixed user roles or group memberships. Rule-based entitlements are evaluated dynamically against a user's profile; they can include a time component, as well as session parameters and requests for specific Web pages. Rule-based entitlements also reduce administration overhead by eliminating manual changes to access privileges when a user's role changes.

The appropriate portal architecture for interaction is one that can help reduce complexity by separating presentation logic from underlying business processes. The interaction of the user with the portal can be represented as a set of events and actions, expressed in XML format and defined using a point-and-click graphical tool. This architecture simplifies the development and administration of portals by ensuring consistent business processes and providing a visual, reusable inventory of the business logic for the portal. A Web flow-based architecture can also be used to enable inter-portlet communication, such as among a set of portlets that supply customer and order information to a portal. For example, entering a student ID number in one portlet causes the display of current class schedule in another and financial aid status details in a third. Such inter-portlet communication goes beyond simply exposing content views to providing an interactive window into applications.

Separating navigation from presentation is also valuable for delivering applications to multiple devices. For example, an e-business checkout process for use on a mobile device requires a more concise sequence of steps than would be used for an online shopping cart checkout.

Integration Services

Integration services are the mechanisms that enable the portal to fit into an existing IT environment. These services should be standards-based to reduce portal integration costs and to leverage Web Services [see Future Trends section] for application integration.

A unified user profile gives an application a single view of the user—be it student, alumni, professor, staff or supplier—across multiple data sources. By aggregating user profiles from multiple data sources including LDAP stores and existing user databases or legacy applications, along with a built-in user profile stored in a relational database, organizations can further enhance the user's portal experience. Based on the values of attributes in the unified user profile, business rules can be defined and used to personalize the interaction with the user.

Pipeline components, sometimes referred to as business logic objects, can be aggregated into an inventory of reusable business logic components. These objects encapsulate discrete units of business logic, which execute narrowly focused processes and application logic. This approach allows components to be added, removed or replaced at any time without affecting the application code. For example, if the payment authorization component is replaced in the sequence of events for committing an order, even though the ordering process now operates differently, no changes to the application code are necessary.

These components can also facilitate process-level communication and data flow between Web applications and enterprise systems using standards like Web

Services Description Language (WSDL) and SOAP, leveraging Web Services to deliver integrated and personalized applications to end users. This integration service also enables and simplifies Web application-to-application connectivity through firewalls for B2B applications.

FUTURE TRENDS

Web Services

Web Services and enterprise information portals are a natural combination, and the intersection of these two technologies has the potential to simplify portal deployment and ease application distribution (Moore, 2001). Because they are self-describing and self-discovering, Web Services simplify portal integration challenges by eliminating the need for hard-coded links from the portlet to each back-end resource. Web Services also facilitate the integration of heterogeneous resources—one of the principal goals of a portal—such as searching other repositories or authenticating users on other systems. In this way, Web Services offer a mechanism to capture content as it becomes available, keeping content current as well as reducing integration overhead.

Portals have the essential user-identity information that provides context for Web Services, context that personalizes Web Services to each user's specific need. Through its authorization and workflow management capabilities, the portal can control access to Web Services.

Publication of portal functions as Web Services in a common Universal Description, Discovery and Integration (UDDI) directory will foster interoperability between portals, vastly expanding the potential value of an enterprise portal. Portals will also help drive initial adoption of Web Services within an organization as vendors design their portlets to be accessed by Web Services. As the proposed standards from the Organization for the Advancement of Structured Information Standards (http://www.oasis-open.org) and Java Community Process help define the Web Service's portal interface, portlet security, and other specifications, adoption of Web Services will accelerate.

Multi-Channel Access

Constituents of all types of institutions or businesses seek access to organizational information that is available 24/7, supports self service and is intelligently managed. Whether the constituent wants access to personal or organizational records from a mobile phone, the Web or a personal digital assistant, the multi-channel access needs of organizations will vary greatly as online access evolves.

Some organizations require little more than wireless e-mail while others need to extend line-of-business applications.

As the three generations of portal software evolve into an enterprise-wide application infrastructure, demand for wireless and mobile access will grow. One of the principal requirements of wireless support is the layering user interface functions, including the translation of page content, for multiple wireless devices. For example, a sales representative accessing the corporate portal via a PDA requires a different view than a customer checking an order using a cell phone. Nearly all mobile and wireless portal products require expensive custom integration work to deliver access to and control of key IT systems. In the end organizations will rely on portal infrastructure to manage device recognition and leverage the reduced complexity of the portal to enable multi-channel productivity applications, such as e-mail and calendar, and custom applications.

For applications that are not designed for the Web, specialized solutions offer a better alternative than attempting to adapt software that was built for a PC-based browsing experience to a cellular phone or personal digital assistant (PDA)—an approach that promises limited success.

CONCLUSION

Organizational agility today rests on the foundation of information technology agility: how fast the organization can respond to a dynamic environment that is driven by the increasing opportunities and possibilities offered by the Internet. How quickly the organization can interpret customer wants and needs—and reflect this understanding in the information and services it supplies—will determine the success of organizations of every type in today's business environment.

To drive business value with portal software, organizations should first reflect on business issues that frame the portal implementation. Customer interactions have shown that portals are most effective when organizations first have developed an understanding of the audiences that will access business process through the portal, and second, have carefully and thoroughly defined their business processes. Only when an organization has cataloged its resource requirements can the evaluation of software infrastructure and portal products properly begin. Finally, the organization should understand the pay-back model for portal deployment in order to correctly measure return on these investments.

As has been shown, portals are an important mechanism that an organization can use to aggregate information and services for the different audiences that it serves. However, the visible portal is only the tip of the iceberg. Realizing maximum value from portal deployment calls for a scalable portal infrastructure as well as powerful portal applications. The technology that is hidden beneath the user interface and presentation layer is the key to connectivity, extensibility and the ultimate adaptability of the portal to changing environments and needs. A successful portal implementation requires an infrastructure with comprehensive capabilities as well as software modules that can be quickly combined and recombined into portal-delivered applications. These capabilities must be inherent in the platform architecture that supplies the foundation for the portal layers. By building on a third-generation portal foundation, organizations can be confident that they are able to meet today's portal requirements while they build an evolutionary path to the portal technologies of the future.

REFERENCES

Correia, J. (2001). *A Major Shakeout Is Occurring in the Portal Software Market* (Gartner Research Note No. SOFT-WW-DP-0030). Stamford, CT: Gartner, Inc.

Delphi Group. (2001, September). *Enabling the Agile Enterprise*. White paper retrieved September 28, 2001, from http://www.usefulportal.com.

Ericson, J. (2001, November 29). The portal comes of age. *Line56*. Retrieved April 21, 2002, from http://www.line56.com/print/default.asp?ArticleID=3175.

Gillett, F. E. (2001). *Making Enterprise Portals Pay*. Forrester TechStrategy Report, August. Cambridge, MA: Forrester Research, Inc.

Hall, K. & Heffner, R. (2001). *Enterprise portals: Evaluation Criteria*. Cambridge, MA: Giga Information Group, Inc.

Katz, R. N. (2000). It's a bird. It's a plane. It's a ...portal. *EDUCAUSE Quarterly*, 23(3), 10-11.

McDonough, B., Morris, H. (2001, January). *Enterprise Portal Adoption, 2001* (IDC Bulletin No. 23759). Framingham, MA: IDC.

Moore, C. (2001). Take it all with you. *InfoWorld*. (October 26). Retrieved April 25, 2002, from http://www.infoworld.com/articles/fe/xml/01/10/29/011029feportal.xml

Phifer, G. (2001). *How Portals Will Slow the 'Infoflood'* (Gartner Research Note No. COM-13-5377). Stamford, CT: Gartner, Inc.

Root, N. L. (2001). *Portal Servers Rely on App and Integration Servers* Forrester Research *TechRankings Brief*. Cambridge, MA: Forrester Research, Inc.

Warzecha, A. (2001). *Justifying the B2E Portal* (Electronic Business Strategies Publication No. Delta, 1032). Stamford, CT: Meta Group, Inc.

Warzecha, A. (2001) *The Right Portal for the Right Job*. (Electronic Business Strategies Publication No. Delta, 1092). Stamford, CT: Meta Group, Inc.

Chapter XIII

Portal Technology and Architecture: Past, Present and Future

Christopher Etesse
Blackboard Inc., USA

ABSTRACT

When you arrive on a college campus, you often get an immediate sense of the institution's history and priorities. You may pass a football stadium and residential facilities prior to arriving at the academic core of the university. Frequently referred to as "the quad," this core area physically links the academic buildings, library, administrative offices and student activity center. All campus pathways lead to this physical center of campus. As college campuses have become more electronically connected, the campus Internet portal can easily be seen as a virtual quad. From the campus portal each member of the university community may be linked to all campus services and information, instantly. Each individual's view of the portal can easily be tailored to unique as well as common needs and interests. This online campus portal is an extension of the brick and mortar of the university. As such, it provides not only the feel and look of the university, but also a common communication tool for off-campus and distance learning participants in the college community.

One of the primary challenges in learning, especially online learning, is interaction, specifically communication. Most, if not all, higher education learning is geared toward communities of communication. As we've seen in other chapters, portals can provide a cornerstone in creating a sense of community on a campus. This chapter explores the origins of online portals. It outlines the technical history of portals including the first portals, the technology underpinnings of today's portals and the ways in which portals will evolve in the future.

Portals can be defined in two ways. One is by the data that reside within the portal and are aggregated for access by the end user of the portal. The second definition is as a framework for accessing, manipulating and interpreting data. This chapter will discuss both definitions. It will also discuss why the portal's data offerings are integral to the balance of these two definitions and how portal framework(s) will become essential in higher education. This chapter will focus on the educational benefits of the past, present and future of portals.

BACKGROUND

From a technology perspective, what is a portal? For the purposes of this discussion, we will describe a portal in the following three ways: a front-end "dashboard" or user interface for any service, a set of standards used to pull or aggregate information from disparate sources and an administrative framework including a graphical user interface for managing an environment.

A portal as a user interface is defined by the services it contains and by the value it gives end users. A user interface includes code as well as graphical elements that allow an end user to interact with a computer or a specific application running on a computer. The distinctive "start" button in Microsoft Windows XP is an example of one element in a user interface. The terms "user interface" and "graphical user interface" are often used interchangeably today. They are the dashboard of today's software. During the 1980s, however, the term user interface often referred to the command line interface in the MS-DOS or UNIX operating system, whereas "graphical user interface" referred to the Microsoft Windows 3.1 operating system or the Macintosh operating system's way of interacting with the computer through the use of a pointer via a mouse. These first user interfaces allowed applications to be built on top of them, such as a Web browser. These browsers made the evolution of portals possible. They provided the user interface to interact with sites like Yahoo and Excite. A portal is the next generation user interface in education. However, a portal does not necessarily interoperate with only one community system; it can also interrogate and present data from other portal systems such as Campus Pipeline, Microsoft SharePoint, JA-SIG uPortal, as well as administrative systems vendors such as PeopleSoft. It doesn't matter where the information originally existed or exists; the portal should seamlessly handle the movement and presentation of the data.

A portal is also an aggregation of standards that are used to locate and collect data for a user. Standards are specifications that define how communication between two systems occurs. For example, in the education world the IMS specifications (http://www.imsproject.org) help define how content can be moved between learning management systems as well as how portals can be integrated with back-office systems such as student information systems. Following refinement and acceptance, a specification can then become a specification of the Institute of Electrical and Electronics Engineers, Inc. (IEEE) (http://www.ieee.org). Once the IEEE has adopted a specification, the next step is for the International Standards Organization (ISO) (http://www.iso.org) to ratify the specification as a world standard.

An example of the aggregation of standards in a familiar context would be the card catalog systems used by libraries. While most of today's library catalog systems are electronic, in the past the systems served as a aggregators of data about books using one of a number of standards. Perhaps the best known is the Dewey Decimal System. Similarly, an education portal such as Blackboard may pull data from campus information systems (student information system, human resource management system, etc.) and from the Internet, and unify their presentation to end users following information industry standards.

Finally, the portal is an administrative framework for managing the campus information environment. The portal's graphical user interface will provide varying capabilities depending on the portal's purpose, but some common characteristics are:

- Information resources are easy to get to, easy to navigate and the portal provides reasons to return. The portal becomes a destination in and of itself.
- The portal provides dynamic information, based on the user's role as well as the user's preferences.

The portal can be customized by many "levels" of users. More advanced and/or more privileged users can make more modifications to make the portal reflect their preferences.

A portal in an academic environment serves as an aggregator of campus data for specific constituency groups. Portals can be useful to the following constituencies:

- Students interested in their grades, courses, community events, clubs and organizations, and in communicating with others on campus.
- Instructors concerned with grades and in interacting with the student information system, and are interested in information about grant status, tenure status, current events and research relevant to their fields.
- Alumni, staff and prospective students are all distinct groups whose environment and viewpoints will influence the types of data in which they are interested.

The next few sections will explore the stages of portal evolution in higher education. In the past, there were static websites for courses in higher education. The next step was dynamic applications that allowed some interaction, for example allowing students to retrieve their grades. Interactive "pulls" from other back-end systems allowed portals to be seamlessly integrated with the existing infrastructure on campus. Finally, data interrogation, interaction and dynamic merging of the characteristics of users and data will present users with the data they need at the time and location relevant to them.

PAST PORTALS

In 1996, I built and maintained a static website for my own course work as well as the courses I was teaching at the University of Kentucky while pursuing my Master's in Computer Science. Figure 1 shows the entrance page into that website. Along the left side buttons allowed visitors to choose to explore examples of my

Figure 1. Example of a common destination page in higher education, circa 1996. Computer science and engineering faculty were among the first on campus to create websites for themselves as well as destination sites for courses they were teaching.

work or the C++ course I was teaching. Note that there was no authentication for the site at this time—anyone on the Internet could visit and look around.

Based on our earlier definition of a portal, this site, largely static HTML, provided the framework for users to view data aggregated in one area. Figure 2 shows a deeper section of my website. CS270 was the C++ course that I taught in the fall of 1996 and the spring of 1997. Much of the organization of this page should be very familiar to those readers who use a course management system such as Blackboard. From this page, users could retrieve documents, notes and homework assignments as well as check their grades online.

This early portal site did incorporate some dynamic code: a PERL CGI script allowed students enrolled in my class to view their grades online.

The early forerunners of education portals were very static in nature and appeared sometime around the mid 1980s. The earliest were just File Transfer Protocol (FTP) sites that allowed users to download and, in some cases, upload files. Gopher sites then evolved, on which users could search for documents using a helpful text interface. The mid 1990s saw the advent of static websites like my CS270 site above. In all these early instructional information repositories, content was not dynamic based on the user; rather, all users saw the same content. FTP sites were crucial for collaboration among researchers and for providing access to public domain software. Gopher allowed users without UNIX skills to search for and retrieve information across the network of instructional and research institutions that

Figure 2. Example of the destination aspect of an early website on campus. This example shows an all-user view of course-related materials for the CS270 course.

were nodes on the original NSFnet. In the mid-1990s, NSFnet was opened up to commercial entities and became the Internet we know today. In a sense, the first browsers on the Internet were really FTP and Gopher clients. It wasn't until 1993 and the release of NCSA Mosaic 1.0 that "surfing the Web" with a graphical user interface could occur. It was this first graphical browser, in conjunction with early Web server software released by CERN and NCSA, that paved the way for true Internet portal experiences.

The earliest portal on the Internet was Yahoo—started in 1994 by two PhD candidates at Stanford University, David Filo and Jerry Yang. Initially their site was static. It started as a way of organizing and tracking all the sites on the Internet that they enjoyed visiting. As the volume of information available through Yahoo! grew, Filo and Yang began to automate the site by creating scripts and a database in which to store the numerous links. Their framework (the scripts and database along with the user interface) provided a static portal in which users could find sites on the Internet (the data).

Figure 3. Example of early dynamic flexibility added to a Web application, circa 1996. Here the student can enter a secret code to retrieve grades for the CS270 course.

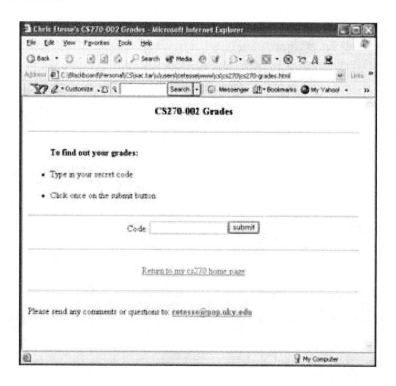

Unlike static pages, dynamic pages take into account who the user is and where he or she is coming from, then create a page on the fly based on available information—often creating this information from a database. The main problem with static websites is that all users see the same information. Because no user-specific information could be provided, authentication wasn't needed to protect the content. Authentication and dynamic presentation were needed. In order to make my CS270 website dynamic and secure, for example, I chose a PERL CGI script that, when presented with a student's "secret code," would parse a comma-delineated file to create on the fly a user-specific HTML page displaying all of the student's grades as well as a grade average (see Figure 3).

CURRENT PORTALS

In late 1996, a group of undergraduates led by Daniel Cane at Cornell University in Ithaca, New York, began to create static websites for their instructors. As more faculty learned of the students' services, they received more and more business, and they began to create tools to assist in the maintenance and creation of the websites. What they soon realized was that the tools could become a software product that would dynamically produce Web pages for users based on the user's contextual information. The undergraduates formed a company called CourseInfo that today is Blackboard Inc.

As these and other programmers began to build dynamic applications, it became apparent that such applications were going to be expensive to maintain

Figure 4. Example of World Class Learning course site, circa 1997. This was one of the first dynamic portals for a college campus. It was hosted by Thompson Publishing and was not available for local installation on campuses.

because there weren't clear divisions between the user interface and the application logic. At this time, the application logic was embedded within the user interface code, the HTML. I built one such example of this early type of Web application for Thomson Publishing. Called "World Class Learning," the application allowed adopters of textbooks produced by Thomson publishing concerns to create dynamic portals for their students.

Figure 4 shows what the syllabus section of a World Class Learning portal looked like. At this stage in the evolution of higher education portals, there were really only three types of users: students, faculty and administrators.

CourseInfo version 1.0 was a course management system much like World Class Learning, but years ahead of it in terms of ease of use and functionality offered. CourseInfo was developed on top of the Linux operating system using PERL CGI's and the MySQL relational database. At the time of CourseInfo's creation, and to this day, PERL was more widely used than server-side JavaScript or Cold Fusion for Web application development. Figure 5 shows an example of the CourseInfo main screen.

Figure 6 shows an example of what the inside of a course looked like in CourseInfo. This screen was dynamically created for s particular type of user. The data were protected behind a built-in authentication system requiring each user of the system to log into the system using his or her own username and password.

Web application development tools based on scripting languages are still on the market. Cold Fusion is one example. Despite their reliance on an older generation of tools, Cold Fusion and Netscape's Enterprise Server before it, enabled a dramatic step forward from the static pages of the past. This period also marked the entrance of two additional components from the early stages of portals:

Figure 5. Example of the CourseInfo main screen.

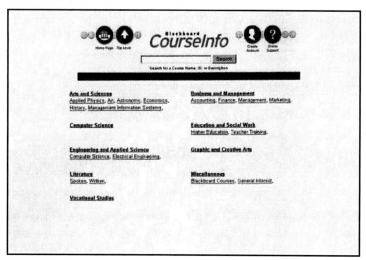

Figure 6. Example of what the inside of a course from CourseInfo.

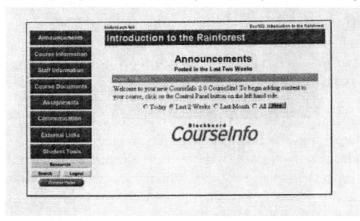

the potential for a portal to offer content from multiple providers and the birth of portal specifications for the online education industry.

In 1996 and 1997, it was uncommon to see online courses from diverse providers aggregated into a single catalog. Instead, each course was located at its own unique URL. The advent of dynamic Web applications enabled single-sign-on authentication as well as the ability for a portal to host content from multiple content providers within a single framework. CourseInfo, mentioned above, could host multiple professors and their courses—and thus was the first broadly deployed instructional portal on campuses. However, the lack of standards for content and interoperability was a challenge for these dynamic websites. This dilemma led to a consensus in the market that standards and specifications for the online learning industry were necessary.

In 1996, Matthew Pittinsky and Michael Chasen were consultants for KPMG's higher education practice in Washington, DC. They noticed a trend in education, namely that almost of the online materials being produced were more a reflection of the information technology infrastructure of the institution than of its teaching and learning mission. At about the same time, CAUSE (now EDUCAUSE),[1] a leading professional society for information technologists within the higher education industry, announced support for an initiative to develop an Instructional Management Standard (IMS).[2] IMS aims to develop open standards and specifications that will ensure interoperability of courseware and course management products in the educational space. Matthew Pittinsky and Michael Chasen spun out of KPMG to create Blackboard LLC, and became the first contractors for the IMS project. The first and most important IMS specification, in terms of portals and technology, was the IMS Enterprise Specification.

The IMS Enterprise Specification[3] defined a set of XML objects and attributes so that various systems on a campus could interoperate. These were:

- A portal and a student information system
- A course management system and a student information system
- A portal and a course management system

The Enterprise Specification was crafted by Blackboard, PeopleSoft and the US Department of Labor. It was ratified in 1999 and adopted by Blackboard the same year.

The challenges at this stage in the evolution of portal technologies were a result of the successes. More and more instructors and departments wanted to be online. The opportunity was clearly to create enterprise portal software that could support an entire campus as a mission-critical system 24 hours a day, seven days a week.

TODAY

At the University of Tennessee, students today can log into http://online.utk.edu/ and interact with their courses, correspond with their professors and participate in the campus community.[4] All of this is provided by the Blackboard system and customizations to it made by the university. This site is the dashboard for University of Tennessee students, faculty and staff.

Today's higher education campuses have complex information environments; portals help to organize and present data from various systems in ways tailored to a variety of distinct user groups. Such modern portals require multiple authentication schemes, data integration with multiple back-end systems, flexible rights and roles, tools with which to integrate other portals and flexible expansion capabilities, not to mention the data and all their various sources. This section will address the challenges colleges and universities face when designing portals to manage their complex environments. We will also look at one portal product on the market today, the Blackboard 6 Community Portal System, which meets the challenges outlined to this point.

With common tools for data integration and authentication integration within a given campus infrastructure, it is possible for portals to contain other portals. Data integration provides value to the campus by allowing it to electronically populate its portal with data from a variety of the institution's back-office systems. Authentication integration allows users to move from one system on campus to another without the need to log in to each separately. Some tools often used for authentication integration include Microsoft Active Directory, Microsoft Passport, Internet2 Shibboleth, lightweight directory access protocol (LDAP), LDAP over secure socket layer (SSL), Kerberos, etc.

So far we've seen the need for authentication and data integration, rights and roles for a flexible system as well as the ability to have portals within portals; the next step in the customization of a portal is its ability to be expanded. The Blackboard Building Blocks program allows for the customization of the Blackboard 6 Community Portal platform as well as the ability to tie other portals into our platform. For example, an administrator logged into a Blackboard 6 Community Portal System might see on the portal page a tab that connects her to the PeopleSoft administrative system portal without requiring her to log in again. Blackboard and PeopleSoft in fact are partners and are presently deploying such an arrangement. Rights and roles defined in the two products give the administrator seamless entrance and exit to and from both systems. A portal is only good if it allows you entrance, draws you back and lets you to move about.

On today's campus, every morning students and instructors log on to their views of the world: their weather, sports and class announcements for the day. In tomorrow's world, this view will be augmented with other technologies, namely small building block applications that will grant portal users unprecedented access to information throughout the university. These applications will exist on the same server, on the same campus or anywhere on the Internet. The glue that will allow these applications to interact with students and instructors on campus is the Blackboard Building Block's program.

Blackboard Building Blocks (B2)—http://buildingblocks.blackboard.com—is an open application architecture that enables the Blackboard platform to easily

Figure 7. A representation of the Building Blocks framework, a key aspect in expanding and customizing a portal. The Blackboard Building Blocks technology and APIs allow the portal to be customized and built onto.

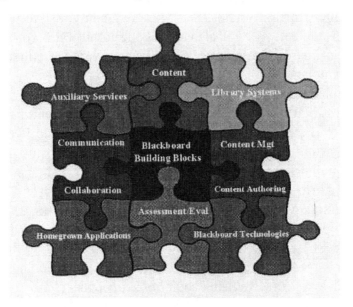

integrate educational, Web-based tools and content that meet institution-specific, discipline-specific or accessibility needs.

Blackboard Building Blocks enables the extension of Blackboard's Learning System, the Community Portal System and the Transaction System. Figure 7 graphically shows how a campus can build on top of the Blackboard platform, extending it by adding or tying in additional tools and content pieces.

An example of integration through Building Blocks would be to tie in another portal, such as PeopleSoft or CampusPipeline, to the Blackboard platform. The same technique could be used to tie the Blackboard Learning System into another portal. CampusPipeline is in fact a Blackboard Building Blocks partner and is doing just that. For example, CampusPipeline uses the Building Blocks APIs to show users the courses they are enrolled in within the Blackboard Learning System as well as aggregating calendar events from within the Blackboard Learning System. The main benefits of the Building Blocks program are:

- Integrating existing technology and infrastructure into the Blackboard platform.
- Incorporating technologies, services, tools or content developed by your institution, other institutions or commercial developers into Blackboard systems.

Figure 8. An example application that ties the Blackboard 5 system to AvantGo for taking content with you on a Pocket PC.

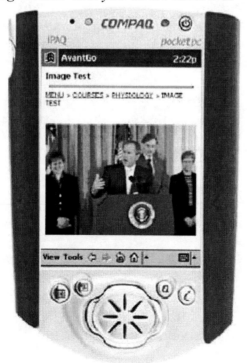

- Providing a supportable framework for external application to be integrated into Blackboard while maintaining the ability to easily upgrade the core Blackboard platform.
- Allowing schools and vendors to plan and develop freely distributed or commercially supported applications as extensions to the Blackboard platform.
- Permitting access to existing Blackboard functionality and graphical interfaces through standard supported APIs.

It is possible to use the Building Blocks program to allow the Blackboard portal and its content to be used in an offline world. Figure 8 shows offline content on the Microsoft Pocket PC operating system.

The ability to take content with you wherever you go provides a large degree of flexibility, but the key component in that ability is the data. If a portal is to be a destination site and data aggregator on a campus, it must have access to the data

Figure 9. An example diagram depicting typical authentication integration on a higher education campus. Having a single source of authentication integration is necessary to implement a portal that can aggregate the data from the multitude of systems while giving the users of the portal seamless access to those systems.

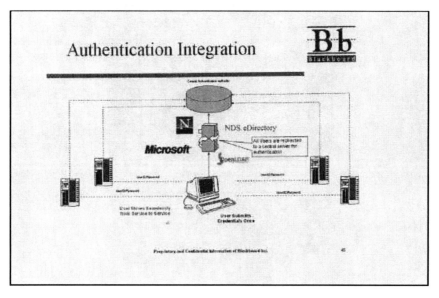

for presentation and allow users to seamlessly interact with those data. The user shouldn't have to authenticate to the system more than once. The Blackboard Community Portal System allows for the integration or replacement of the default Blackboard authentication system with a number of other authentication systems. Integration has been successful with Microsoft's Active Directory and Passport, Novell ActiveDirectory, Kerberos, LDAP and others. Figure 9 diagrams the typical integration on a campus.

Authentication integration allows the portal to pull data from other systems such as other portals, student records systems and the Internet in general. For example, the Blackboard 6 Community Portal System can access and integrate user data from the Blackboard 6 Learning System as well as, for example, two other learning management systems. It could present users with lists of all the courses they were enrolled in. If all three systems used the same authentication system, the users could leave the Blackboard 6 Community Portal System and enter either of the other learning systems without having to re-authenticate. Another key requirement of authentication integration is data integration.

Figure 10 shows a number of ways in which the Blackboard system can integrate into a campus environment of multiple different systems. This is extremely important in designing portal infrastructures for higher education, because the constituency groups are often represented in a multitude of different systems. For

Figure 10. Example of data integration among disparate systems on a campus using the Blackboard Learning System as well as the Blackboard Community Portal System. This diagram depicts the numerous ways of doing integration, one of which is via the IMS Enterprise Specification.

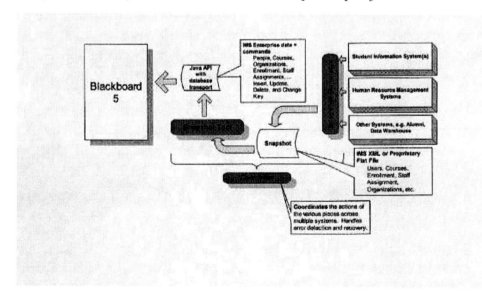

example, students are in the Student Information System (SIS), faculty may be in a separate Human Resource Management System (HRMS) and alumni are in the separate alumni system. For a user to take advantage of authentication integration, data integration must ensure that the user is identified identically in both systems, for example under the same username or the same unique user key (ID).

In a portal, the user's identity is extremely important in determining what they will be presented with; flexible rights and roles are one method of customizing the user interface after we know who the user is. Flexible rights and roles determine who the user is contextually (a student or instructor for an example), where they are and what they have access to. For example, when a student authenticates into the Blackboard 6 Learning System, he is presented with a user interface (portal) specifically tailored towards students. His role definition determines which tabs are displayed, which modules or channels he sees on a particular tab as well the content that appears within specific modules, such as the courses he is enrolled in. Module and channel are words used interchangeably to define the discrete areas on a portal page that provide a particular type of content. Finally, rights and roles are also important in determining what users encounter within a portal and specifically whether users encounter other portals within them.

The Blackboard Community Portal System allows for data to be aggregated from the Blackboard Learning System through direct interaction with Building Blocks APIs in order to show users all the information they are interested in:
- The courses they are teaching or enrolled in.
- Announcements for their courses.
- Calendar events for themselves, the system and specific courses.
- Their tasks.

Items such as those listed above are the key data education users will return to and return to often. But, the Blackboard Community Portal System also pulls and can publish data from/to multiple other systems using various technologies. Examples include:
- The local weather via HTTP.
- Presenting courses inside the uPortal by publishing RSS channels.
- Presenting Java applets.
- Various data sources via XML.
- Building Blocks for custom extensions to pull in data.
- Microsoft .NET services via SOAP.
- All of the information is dynamic; the weather is updated as the weather changes, calendar events change as new assignments are posted and the course listing adjusts based on add/drop. Out of the box the Blackboard

Community Portal System comes with content from outside sources as well as tie-ins to the Blackboard Learning System.

Today's portals are challenged by their access to data. They are limited in what systems they can pull data from, as well as the ways in which they can present data to the portal viewer. In the future, portals may be able to dynamically fetch data from multiple disparate sources on the fly.

FUTURE STAGE

In the future students may be able to log into their portal pages each morning and encounter a much more dynamic world. They will be presented with data based on their particular place in life, the courses they are studying, the assignments they need to complete as well as information related to their friends. The community-building aspects of future portals may dynamically update their users about the location of a friend or a classmate who may also need to complete a group assignment. The portal will have the logic and tools to make this interaction possible.

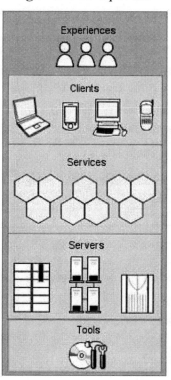

Figure 11. An example of the Microsoft .NET infrastructure. These key infrastructure components will serve as the underlying environment for pulling multiple systems and disparate data together into a portal.

Portals will continue to evolve over the next couple of years, so much so that the frameworks themselves should effectively disappear. We won't recognize portals as being distinct; they'll just exist. A parallel exists now in the way most Microsoft applications on Windows systems look the same; their manufacturers all use Windows Microsoft Foundation Classes components to build the user interfaces. In the future, multiple portals may provide the information experience to the users, but users will not be able to tell the difference as they seamlessly move between them. Some of this is already happening with data and authentication integration today, but it will be more fluid in the future. Portals will also tailor themselves to users with a finer granularity, changing their appearance based on the speed of the users' Internet connections, where users have been and what their tastes are. Vast, user-specific knowledge bases are likely outcomes of today's evolving information environments.

While today's technology is largely based on Java, PERL, ASP and Cold Fusion, future portals will incorporate more of a Web services paradigm. Microsoft's vision of the future includes numerous applications spread across the Internet, all communicating via Web services. Figure 11 shows an example of Microsoft's .NET hierarchy of Web services.

In the future, as a user logs onto a portal, various services across the Internet will be called to create a custom view for the user at that moment. Intelligent caching techniques may be employed to ensure that the data provided are the latest, but more importantly to ensure that there is no delay in loading the portal pages as the underlying portal technology attempts to go out across the net to retrieve the data from all the resources pertinent to the user.

IMS, discussed earlier, is one of many technology standards that will provide the framework for interoperability among portal applications. This interoperability will provide the basis for what portals will become. We've seen that they are destination sites and they are taking the offline world online, but they will also start to enable other applications for users to interact with.

Figure 12. An example of a marketplace application on top of the Blackboard 6 system. This destination site also offers eCommerce functionality for purchasing tickets online and paying parking tickets, as well as making donations to the university.

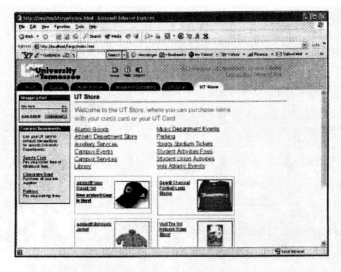

One new application available through a portal is simplified online payment technology. An example is eDebit functionality. With it, the Blackboard Transaction System allows campuses to take current offline payments to the online world. At present, for example, students rent chemistry goggles at the beginning of the term by paying the department secretary in cash. In the future online portal world, students will be able to log in to the campus portal and rent their goggles using the stored value card system on campus. Of course, they'll need to physically pick the goggles up in person! Figure 12 shows an example of what an online marketplace might look like in a portal.

CONCLUSION

Portals will not be bounded by limitations of human communication. They are becoming global centers in education, uniting academics and community with a unified interface. From the days of Yahoo and CourseInfo, portals have always been destination sites. Today and in the future, portals will serve as aggregators of education-related data and the foci of online communities, continuing to entice users back for more. We've seen static websites pave the way for the dynamic applications of today with constantly updated information. Authentication and user roles have allowed for dynamic customizations to the user interface or dashboard, allowing administrators to require certain data but also giving users flexibility in the data they want to view and how they wish to view it. We've seen how the Blackboard Community Portal System provides a wide variety of tools and customizations to make the portal a destination site for a campus. And we've talked briefly about where portals may take us in the future with intelligent Web services, and commerce options. By providing the global framework for conducting education online, portals serve as aggregators for user data; but as this body of data becomes ever more complex, advances in portal technology will enable a much more personalized approach, enriching the educational experience for students, faculty and administrators alike.

ENDNOTES

[1] http://www.educause.org
[2] http://www.imsproject.org
[3] http://www.imsproject.org/enterprise/index.html

Chapter XIV

Building a Virtual Campus

Stephen Ast and Cassandra Gerfen
eCollege, USA

ABSTRACT

For the higher education world, comprising three million faculty and administrators, 15 million students and 60 million alumni (Budzynski & Zabora, 2000), a Web presence is essential. But the means by which institutions "go online" can differ dramatically, including the implementation of brochure websites, development of online registration systems, offering Web-based course supplements and distance learning courses, putting administrative functions online, and giving students Web access to extracurricular resources and other networked information. When delivered together, these functions represent a comprehensive platform, while alone they create a fragmented Web presence.

In this chapter, we will explore the trends leading up to the need for an institution-wide solution; how the eCollege CampusPortal[SM] can connect a campus' administrative, academic and community aspects together through a seamless, single point of contact; the development process and technology that makes this possible; and the building of a virtual campus at Montana State University-Billings.

Copyright © 2003, Idea Group Inc. Copying or distributing in print or electronic forms without written permission of Idea Group Inc. is prohibited.

EXPLORING TRENDS LEADING TO AN INSTITUTION-WIDE SOLUTION

The Internet Becomes Part of Everyday Life

The Internet has become part of our daily lives for everything from communicating, shopping and accessing entertainment, to investing and operating a business. The number of Internet users is expected to reach 320 million worldwide by 2002 (Webber & Boggs, 2001), and for this group, checking e-mail, retrieving news and even analyzing purchase decisions online could become routine. The Internet has also become a primary means for students to conduct research, especially when it comes to pursuing and "virtually" touring prospective institutions.

As more and more prospective students started turning to the Internet to help them decide which institution would best meet their educational objectives, institutions began building brochure websites to market to these students. Brochure websites serve the immediate need of familiarizing students with campus life through text and photos, but are often limited in terms of interaction and administrative functionality.

eLearning Takes Off

The Internet has transformed nearly every facet of life, including education. Both in the classroom and from a distance, this medium has been deemed a key resource to make education more accessible, engaging and interactive. The term "eLearning" was coined to define this practice of forming key learning relationships and processes through the use of Internet technologies.

eLearning is becoming more prevalent in one way or another at many institutions, ranging from those that simply accept applications online, to those offering hybrid courses that meet both in the classroom and online, to those delivering full online degree programs. International Data Corporation (IDC) estimates that schools offering eLearning will double through 2004, resulting in nearly 90% of all higher education institutions offering some type of eLearning by that year (Webber & Boggs, 2001).

The push toward this new way of learning has resulted in faculty members implementing a variety of eLearning tools in their individual classrooms, departments building their own websites and online distance programs, and institutions developing strategies for online administration. This creates an inconsistent approach to eLearning, as varying tools and platforms are being used across campus to fulfill individual eLearning objectives.

Students Demand Technology

Higher education enrollments are on the rise due to a record number of high school graduates, a greater percentage of students attending college and a record number of working adults returning to college (Budzynski & Zabora, 2000). With the surge of nontraditional adult learners and the tech-savvy 'Net Generation encompassing the 30% of our population born after 1977 (Tapscott, 1998), greater strain is being placed on existing bricks-and-mortar facilities. Further, today's students are demanding that technology-based learning and administrative processes be integral parts of their educational experience.

According to a recent study conducted by eBrain Market Research, more than 65% of U.S. adults in online households are interested in continuing their education via distance learning (Consumer Electronics Association, 2001). As we experience a shift toward a knowledge-based economy, demand for lifelong learning and training opportunities increases and it becomes more important than ever for institutions to keep strong ties with their alumni.

Couple this group of lifelong learners with traditional students graduating from high school in an era when 98% of public schools are connected to the Internet (Cattagni & Westat, 2001), and the need for wired campuses becomes even greater. In fact, the 'Net Generation is selecting colleges in part based on how "wired" they are, ultimately giving better-connected campuses a competitive advantage in attracting students. Recent high school graduates are confident daily Web users, expecting Internet-based resources to be as much a part of their education as time spent in the physical classroom.

Today's technology-literate student population is motivated to use the tools available to them, even if their instructors and administrators are not. Individual students and student organizations have taken it upon themselves to develop autonomous Websites for their various clubs and teams, which may or may not be consistent with the institution's culture.

The Result: A Fragmented Web Presence

Each of these trends has contributed to a fragmented approach to eLearning, as evidenced in the lack of integration many institutions experience among their various Web assets. Students, faculty and administrators are all trying to leverage the Internet in education, and although they are moving in the right direction, concerns that a disconnected Web presence will negatively impact the institution's image and adversely represent the quality of the institution online are surfacing. Additionally, because faculty and students often become confused when trying to learn to use services delivered on multiple platforms, they are spending additional time on the technology, rather than focusing their efforts on the learning material.

There exists a compelling need for an institution-wide solution through which administrators can streamline processes; students and faculty can access academic, administrative, social and personal aspects of the campus through a central location;

alumni can stay in touch with the institution and vice versa; and the entire campus community is brought together to interact online. A portal environment, ranked together with search engines and online communities as the most popular Websites visited by U.S. Internet users (Shop.org, n.d.), enables institutions to bring together its constituents through a single point of contact.

Of the five major trends that will transform the landscape of the education and training industry over the next decade, education portals will play a key role in providing increasing eLearning opportunities for the education community.—Merrill Lynch (Moe, Bailey & Lau, 1999)

However, the portal must be more than a point of entry to keep students, faculty and administrators engaged outside of the bricks-and-mortar environment they are accustomed to—it must combine the institution's eLearning and eCommunity functions seamlessly. Created from a desire to provide everything a student can do on campus online, the eCollege CampusPortal comprises the tools and technology to take an institution's Web presence to this next level.

ENVISIONING AN INSTITUTION-WIDE SOLUTION

eCollege—a leading provider of eLearning software and services to the higher education market—designs, builds and supports high-quality online degree, certificate/diploma and professional development programs for colleges, universities, school districts and state departments of education. As a pioneer in the industry, eCollege built the first online campus in 1996 with one course serving just four students. As of the end of 2001, the company had built and hosted more campuses than any other provider in the industry, supporting approximately 300,000 student enrollments.

At its founding, eCollege envisioned a day when all services a student could access on campus would be available online, from applying for admission, registering for courses, and accessing financial aid information and career counseling, to taking courses, purchasing textbooks, communicating with peers and professors, conducting research and checking grades. In many cases today, this information is delivered through one-way communication. Students access the brochure website, read or download the necessary information and then send an e-mail with follow-up questions or actions; rarely, if at all, do they experience two-way interaction with faculty, administrators and other students.

As eCollege and the eLearning market matured, it became evident that the 90-plus percent of college students who are online today (Budzynski & Zabora, 2000), and institutions in general, were ready for a more interactive approach to campus services. The vision became clear—to build and deliver a single, consistent, seamless environment capable of integrating the academic, administrative and community aspects of the entire physical campus online.

eCollege examined industry statistics with regard to the number of people relying on the Internet in education, their current application of the technology and their demands, and used this data to drive product development efforts to better address the needs of the marketplace. It was also important to recognize that even though we were dealing with a tech-savvy student population, in some instances individual faculty and administrators would not fit this profile. To simply offer the tools to meet the demand would not be enough. The solution needed to accommodate a diverse group of users.

In the fall of 1999, we set out to further develop our CampusSolutions^{SM} product line to include more community-based features, accessible to a wide variety of user groups. The result, CampusPortal, not only provides more interactive tools that allow for two-way communication, but it also combats the trend toward a fragmented Web presence.

CampusPortal complements an institution's physical presence online, replicating many of the interactive and community aspects that students and faculty experience on a traditional campus. The product provides institutions with a unique home—a single place on the Internet that students, faculty and administrators will visit several times a day and depend upon as their primary source of information. It also serves as a resource for prospective students and alumni, who may or may not be visiting as frequently, but still see the portal as a valuable tool for staying abreast of campus information.

Designed to reflect an institution's culture, CampusPortal assists an institution in evolving its one-dimensional campus Website into one that promotes user-centered Internet communication, increases operating efficiency and eliminates organizational boundaries. Specifically, it helps an institution:

- Create more extensive and more consistent communication across campus
- Enhance the overall campus experience and quality academic outcomes
- Strengthen alumni relationships, enabling alumni to stay better connected to their collegiate roots
- Invigorate faculty interaction and collaboration on teaching, academic research and service projects
- Target community groups with specific content
- Contribute more to the personal growth and development of every student
- Increase the effectiveness of student support services
- Provide students with a single, identifiable look and feel for all eLearning resources
- Allow for the tight integration of data between online courses and campus services

CAMPUSPORTAL'S DEVELOPMENT HISTORY
Establishing a Vision and Conceptualizing the Design

A project of this scope could not be undertaken lightly. The decision to build CampusPortal was based on extensive evaluation of customers, competitors and markets. Once this decision was made, a careful design and development process was launched.

Any large development project begins with a vision. During this period, the broad strategic goals of the product are established and the scope of the product features is defined. The vision for CampusPortal was set with the following goals in mind:

- To serve as a gateway for students into their institutions' online programs, delivered through the eCollege platform
- To serve as a gateway for an institution's administrators into the eCollege administrative and course management systems
- To foster eLearning communities by providing areas for interaction outside of the classroom, but still within an academic environment
- To serve as an academic and community portal, by providing access to academic resources such as study groups, tutoring services and online study guides; campus information and events; and general interest resources such as news headlines, stock portfolios, Web search engines and online vendors
- To offer campus administrators a centralized location from which to reach multiple constituent groups
- To carry strong branding opportunities that allow users to recognize the CampusPortal as a communication tool of the institution it represents
- To provide a complete, outsourced solution for higher education institutions striving to provide superior eLearning programs

Once the goals were defined, the next step was to initiate the *Conceptual Design* of CampusPortal. This exercise required eCollege's team to develop a complete and detailed set of requirements for the product. We began to look at high-level strategic directives and translate them into blueprints, which would be used to build the platform. Extensive conversations with potential users of the portal ensued, confirming our strategic goals for the product and helping us to further define the feature set.

With the definition phase complete, our team was ready to tackle the question of whether to build or buy: was this a software solution that eCollege should build from the ground up, implement using some third-party components or acquire as a third-party turnkey solution? eCollege carefully considered the scope of the project based on the business case before making the buy vs. build decision, looking indepth at a number of existing portal products in order to understand their functionality, scalability and competitive advantages. We then looked at integration opportunities and conducted research on third-party products that met some of our specific feature needs. Although this was a rapidly growing marketplace with many companies vying for market share and eager for relationships with our application service providers, partnering with a third-party vendor would require indepth investigation into the long-term capacity and financial stability of the company. It would also add a level of complexity to our existing customer relationships as we approached implementation stages.

In reviewing potential third-party vendor relationships, we had to consider the issue of accountability. If a technical problem should occur within the product, would the user call eCollege or the third-party vendor? Relationships such as this can become confusing to the end user. We were concerned about relying on other companies to match our standards, given our overall course management platform systems availability (2001) of 99.97%—unrivaled performance in the eLearning industry. Any relationship with a third-party provider would have to include a Service Level Agreement and diligent training program to enable eCollege technical support staff to continue to provide high-quality service to our users.

At the time we were reviewing our options for CampusPortal, the trend in the industry was to give products away and recoup costs through advertising revenue. Most third-party vendors that we evaluated proposed this business model to eCollege in order to gain our business. However, because of the unique nature of the higher education market, eCollege reviewed this option very closely and with skepticism—higher education institutions typically shy away from commercializing their offerings, especially in the academic arena. Since one of our main principles for CampusPortal was to provide an eLearning community, it did not seem viable to force institutions to display advertising as payment for the product. It is also important to note that many portal companies relying on an advertising-eCommerce model have since gone out of business.

Ultimately, we concluded that existing portal vendors did not have the required features and flexibility, nor could they reach the level of integration with existing eCollege products that we had identified as necessary for the success of the product. The decision was made and we focused our efforts on internally building an integrated eLearning community for higher education institutions.

Logical and Physical Design

The goal of logical and physical design is to arrive at a software solution—first an abstract object model, then an implementation strategy—that best meets the core user needs and feature set defined in the Conceptual Design phase.

From the outset, we knew that CampusPortal, like all eCollege products, would function as a hosted application, integrate with our existing products/system and be highly scalable. As an eLearning community portal, the product would have to accommodate a user pool well beyond those currently enrolled in coursework, extending to alumni, prospective students, faculty and staff. These considerations dictated an "n-tier" architecture involving a clear, logical (and potentially physical) separation between the presentation layer, the business layer and the data layer.

Beyond this point, the design needs of CampusPortal diverged significantly from those of eCollege's then-current eLearning systems. As an academic community portal, the new product needed to accommodate heterogeneous content from a wide variety of sources, including third-party vendors that might change their content on a daily or hourly basis, content published and edited (perhaps with great frequency) by school administrators, and content driven by an institution's own back-office administrative systems. Above all, this content had to be *targetable*—not just to the users of a given institution, but to individuals and groups within that institution's community based on any number of roles or affiliations.

Technical Design
Wireframe and Nuggets

While many portal products use channels to draw content from external resources, the eCollege CampusPortal addresses content management in its own unique way. CampusPortal uses nuggets, which represent content containers that are editable by a system administrator. Nuggets consist of independent units of content generated as HTML streams by their own dedicated system components. Instead of linking to an external resource for content population within CampusPortal, content within each nugget is housed in the eCollege database. To meet its requirement of targeting specific user groups, eCollege combined these content nuggets with a wireframe display, a presentation tier page whose sole function is to marshal the appropriate nuggets for display to the designated users.

Nugget components can acquire their content from any source within or outside the system:
- they can be very "thin" (for example, containing hard-coded HTML content or links to external Websites); or
- they can contain complex logic for manipulating or rendering data for display.

In its first release, CampusPortal featured nugget components that displayed dynamic third-party content; components that displayed hard-coded content and features; components that displayed client-specific, fully editable content in a variety of different formats; and components that provided gateways to other eCollege products. This range of nugget types is expected to grow as eCollege addresses future needs of clients and the academic marketplace. The only static rules for nugget component implementation are that they support the business tier operating environment, that they have a public interface conforming to the platform's application programming interface (API) and that they function independently of any other nugget.

The requirement of nuggets functioning independently was crucial to our overall design goals. The instance of a given nugget must be able to display, by itself or with any number of other nuggets, on any tab-page of any school's CampusPortal, to any number of users, in any order.

While the task of collecting and formatting content was assigned to the nugget objects themselves, the task of deciding whether, or where, a given nugget would display was left to the wireframe, based on roles/rights information entered into the database when that particular nugget is published. Publishing a nugget means associating a specific instance of content generated by a nugget component with a specific set of user roles for a specific campus—typically roles such as campus administrator, faculty member or student, though these can be made much more granular. In fact, individual schools are allowed wide latitude in the array of roles they choose to create for their CampusPortals.

The publishing process also defines on which tab/page of the wireframe a given nugget instance will appear, assuming the user has appropriate rights to view the nugget. Since users must authenticate before accessing CampusPortal, the wireframe knows the user's identity and role(s), and can therefore decide which nuggets to display to that user, drawing from the whole array of nuggets that *could* be displayed on a given page.

CampusPortal is made up of seven persistent pages, or tabs, that can be turned on or off based on an institution's requirements (see Figure 1).

Figure 1.

Tab	Features
Home	Search, personalized web links, advertising support, announcements, custom content and an Administrative link for users with appropriate permissions.
Academics	Announcements, online tutoring, study groups, online study guides and research tools. For students taking classes within the eCollege System℠ the Academics tab also includes direct links to their online courses.
Community	Support for announcements, clubs, a student union and a faculty conference center.
Services	Academic Services, Campus Services and any other custom links you want to include. *(for example, a link to department pages)*
Web	Announcements, News Headlines, Personal Stock Portfolio and Web search.
Alumni	Alumni chapters and alumni services.
Marketplace	Online Yellow Pages and links to multiple online vendors.

CampusPortal℠ Page Features

User identity and role information is passed into the nugget components themselves so that the latter, if necessary, can make more specific decisions about content display. For example, while some nugget objects (including many of the third-party integrators) would display their content identically for all published roles, other nuggets (for example, a student's course list, list of clubs joined or favorite websites) would display content unique to each user. The editable nugget types were a special case, displaying content to all users, but an edit button only to those users whose roles authorized them to edit.

Editable Nuggets and Nugget Editors

Editable nuggets are central to the product's function as a campus communication tool, and as such they absorbed a good deal of our design and development efforts. Three main formats of editable nuggets are:
- *Announcements*, in which a single nugget contains one or many separate rich-text paragraphs, each with a start and end date for display;
- *Highlight*, in which individual annotations and links to other documents appear in a one-, two- or three-column format; and
- *Topic Box*, in which multiple document links appear in an expandable tree menu. The header logo and footer boilerplate text are also editable.

In each case, the content had to be editable without disturbing the complex Javascript and HTML necessary to generate the proper formatting for that nugget type. Moreover, every editable nugget, regardless of its specific content, had to conform to the style and color "branding" that the particular institution chose for CampusPortal. Of course, as long as a nugget adheres to its basic interface requirements, new types can be added easily, leaving the door open to the establishment of new formats in future releases.

To meet this challenge, eCollege segmented the content-and-presentation mechanism into three layers. Editable content for a given nugget instance is stored as XML in physical files; at runtime this unique XML content is converted into properly formatted HTML by means of an XSLT template, one for each of the editable-nugget types (or actually two, one for Netscape-compatible HTML and the other for Microsoft Internet Explorer-compatible). The HTML generated by these, as well as all other nugget types, was designed so that a single cascading style sheet (CSS) could be applied to all content elements on a given partner's site. The CSS itself is generated uniquely for each institution through the Style Manager, a separate administrative interface that allows font and background color to be selected for more than 20 different content elements.

The nugget editors also represented a significant design and development effort. For ease of use, it was decided that the editing interface, essentially a free-

Figure 2.

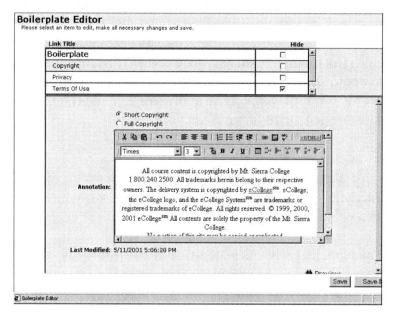

standing Web application in its own window, would be invoked directly from the nugget to be edited; and while the different nugget formats dictated some differences in the graphic user interface (GUI), eCollege kept the overall interface of the editors as uniform as possible (see Figure 2). While many of the functions within the editors are inherent to most word processing programs, the eCollege Visual Editor allows users to switch between the "design" view (shown in Figure 2) and the HTML view that allows users to employ more advanced techniques if desired. Other functions are consistent to word processing programs allowing users to link in images, websites or documents; change text size and color; align text; insert tables and many more basic functions.

Content Caching

The flexibility of the n-tier approach and its benefit to the overall CampusPortal design is visible in the content caching technology. Designers recognized that in a CampusPortal with a large number of dynamic nuggets whose content sources varied from high performance database back-ends to dependencies on third-party websites, scalability and performance could become issues. The solution was to completely separate the process of role-based content display at runtime (and the data store supporting this) from the process of content generation itself. All of the nugget components would still be used to generate HTML content, drawing their data as before from XML files, third-party sites or other locations. However, whenever possible, these HTML streams would be cached in the system's SQL database, each record associated with a given nugget instance, so that the wireframe could simply draw from this cache in order to display a given array of nuggets, rather than calling the nugget component itself. With the cache in place, the wireframe would only have to call on the nugget component to (re)generate the content if it detected that the cached version was not yet present or had expired.

Several additional elements of the caching technology are worthy of note. The cache exposes an API that allows any nugget to clear itself from the cache. This is typically used so a nugget can clean up after itself when edited. Cache expiration can be set on a per-nugget basis, providing an additional level of flexibility. Finally, nuggets can be excluded from the cache in the event of a business scenario in which the content needed to be fetched directly from the source every time it was requested.

The caching technology slides neatly into CampusPortal, between the wireframe tier and the nugget tier. This is testimony to the advantages inherent in two basic principles our team followed in the design of CampusPortal as well as other products: n-tier architecture, which dictated the clear separation of presentation mechanisms, business logic and data store(s); and object-oriented component design, which drove encapsulation of functionally distinct code in different business objects. As our experience with CampusPortal demonstrated, up-front investment in sound architecture can yield very tangible benefits in product performance and delivery time. This approach, including encapsulation and the isolation of logical tiers, will continue to drive CampusPortal's technological success well into the future.

IMPLEMENTATION AND ADMINISTRATION OF CAMPUSPORTAL

Once a product decision has been made, the implementation and subsequent administration and management of the portal must occur. Unfortunately, institutions rarely dedicate enough qualified resources to these make-or-break processes, either on the technical side or the administrative side, to make the portal successful. In many situations, the portal never reaches its full potential or, even worse, never launches. To avoid this pitfall, eCollege designed CampusPortal to be easy to implement, modify and customize, to meet the needs of the marketplace.

CampusPortal was successfully released through a beta rollout in the spring and summer of 2001. The following is the story of Montana State University-Billings (MSU-B), which participated in the beta process, and its implementation of the CampusPortal product. The remainder of the chapter focuses on the

Figure 3.

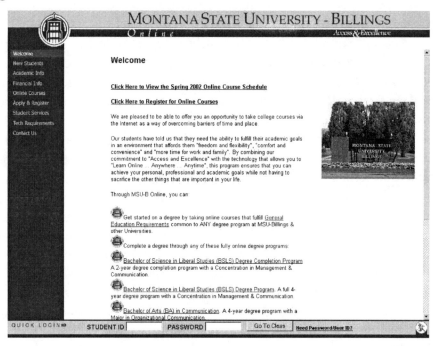

institution's goal of elevating its virtual campus for distance learners (see Figure 3) to a more collaborative platform, and the rollout approach eCollege used to make MSU-B's portal dream a reality.

History

MSU-B Online is a tremendously successful online distance learning program, with 3,810 student enrollments through the fall 2001 semester (see Figure 4). The following programs are available entirely online:
- Associate of Science Degree Program
- Bachelor of Science in Liberal Studies "2+2" Degree Completion Program
- Bachelor of Science in Liberal Studies (BSLS) Degree Program
- Bachelor of Arts in Communication Degree Program
- Bachelor of Applied Science (BAS) Degree Program
- Master of Arts in Public Relations Degree Program

These degree programs are built on the premise that interaction among students, peers and faculty leads to a quality learning experience. MSU-B wanted to create a campus community, allowing students to interact with each other and the school's faculty and administration via the Web. MSU-B Online's current student population is over 75% female, and 65% are adult learners (between 25-55 years

Figure 4.

Student Growth by Course Enrollments				
Enrollments	Summer	Fall	Spring	Annual
1998-1999	n/a	36	108	144
1999-2000	48	153	265	466
2000-2001	230	515	742	1487
2001-2002	637	1076	1300*	3013

old). Over 85% of all online students are degree seeking, and 79% say they wouldn't be able to complete their degrees without the online courses and support services (Lacy, 2001).

Goals

Fulfilling the following key goals would enable MSU-B to offer an eLearning and academic portal for its online students:
- Implement CampusPortal for the fall 2001 term for all distance learning students
- Allow students to register online for eLearning courses
- Train students, faculty and staff on the functionality and features of CampusPortal
- Provide school-based content (text, forms, etc.) and Web-based resources (tutoring service, study guides, news headlines, stock quotes, shopping)
- Brand CampusPortal as the official site of the MSU-B distance learning programs, and make it visually appealing and easy to navigate
- Offer support services that enable students to have a successful learning experience
- Provide faculty with the ability to easily customize and manage content without knowledge of programming languages

Implementation Process

Having less than two months to implement the MSU-B portal left little time for committee meetings. Mr. Kirk Lacy, Director of MSU-B Online, came to the eCollege headquarters in Denver, Colorado for two days of intensive meetings that focused on matching portal functionality with MSU-B's goals, as well as developing a phased rollout approach so that both eCollege and MSU-B could prioritize the workload and meet the primary goal of launching in the fall of 2001.

An action plan, developed by eCollege during previous portal implementations and modified for the MSU-B implementation, was put into place. During the visit, eCollege and MSU-B representatives were able to:
- Discuss and confirm MSU-B's portal goals
- Establish the core features/functionality of the MSU-B portal
- Define and create portal user groups/roles
- Determine content for each page and each information element ("nugget")
- Assign content to each user group
- Provide *Visual Editor* training, including inputting content into new topic boxes; highlighting links on custom content pages, external websites, existing files; and the ability to view the different content layout options

To assist in defining which nuggets needed to be viewable and/or editable for each role, a spreadsheet was created to define the tabs, containers (location on the page), and pre-published nuggets. MSU-B then added rows to address additional nuggets, and columns to add more user roles (see Figure 5).

By the time the implementation visit was complete, the portal shell was created and the focus then moved from portal configuration to developing and inputting content into the portal. At this point, although the bulk of the work shifted to MSU-B to decide exactly what content would be included, eCollege continued to provide consulting and Help Desk support to ensure the project's success. Content development is the most time-consuming aspect of building a portal. MSUB had the advantage of implementing other large projects in a short timeframe, and thus was able to identify and assign qualified resources to get the portal built quickly and correctly. Also, because MSU-B and eCollege had worked together extensively in the past, a good working relationship had been established, and continued. Without MSUB's experience with large undertakings with eCollege, this project would not have been completed as quickly or thoroughly.

Figure 5.

Tab	Nugget	Student	Faculty	Administrator	Other
Home	Welcome Message	View	View	View/Edit	TBD*
Home	Events Calendar	View	View	View/Edit	TBD*
Home	Web Links	View	View	View/Edit	TBD*
Academics	Enrolled Courses	View	View	View/Edit	TBD*

Nugget/Role Manager

When the content development was nearing completion, MSU-B assembled a test group made up of students, faculty and administrators and had them test the site. The goals of this group were to:
- Collect feedback on the CampusPortal layout and content
- Test the content on multiple browsers (IE and Netscape), browser versions and operating systems
- Enable more eyes to review the site as part of the quality assurance testing

The full portal launch, scheduled for the following week, wouldn't have been nearly as successful without the input from this group. This experience served as proof of our belief that all new releases, whether new version code or modifications and implementations made by an institution, should go through a quality assurance process. Making mistakes in front of a test group is expected, and generally viewed as a positive; making the same mistakes in front of a real, and usually much larger, audience is unacceptable. Additionally, by involving this diverse group in the testing process, MSU-B saw a faster rate of adoption because the users felt a sense of ownership in the finished product.

Looking Forward

With the MSU-B portal (see Figure 6) officially launched, both parties had a chance to look at the bigger picture. Future goals for the MSU-B portal that have been identified include:
- Integration with MSU-B's Student Information System to transfer enrollment data, grades, etc., in near real-time
- Adding additional content for support services
- Offering the portal to on-campus MSU-B students as their homepage
- Developing more specific user roles, including prospective students, alumni and staff
- Customizing content based on the user's degree program
- Decentralizing the control and responsibility for all portal content

Not surprisingly, these goals align directly with eCollege's goal of its offering Campus Solutions[SM] that ensure greater and consistent communications across campuses, increase effectiveness of student support services and strengthen the university's relationships with its alumni.

The outcome of the implementation is summed up in the words of director Kirk Lacy. "This comprehensive and dynamic platform enhances our services to our online faculty, students, staff and alumni," he said. "CampusPortal facilitates greater interaction and provides access to a richer online resource center that directly affects the success and satisfaction of our faculty and students. The adoption of CampusPortal is yet another milestone in our evolving partnership with eCollege."

Figure 6.

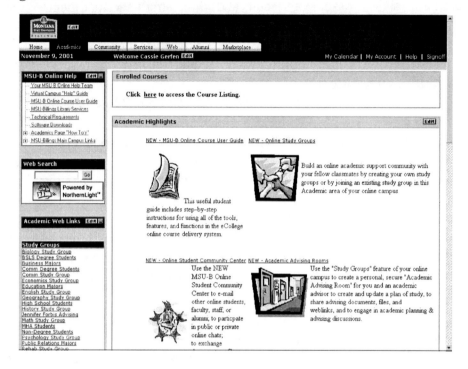

ABOUT eCOLLEGE

eCollege is a leading provider of eLearning software and services to the higher education market. eCollege designs, builds and supports high-quality online degree, certificate/diploma and professional development programs for colleges, universities, school districts and state departments of education. The company provides the technology and services that enable colleges and universities to offer a synchronous and asynchronous learning environment for distance and on-campus learning.

eCollege is leading the way in using technology to make education more accessible, engaging and interactive—from online campuses, to online supplements for traditional courses, to full online distance programs. The company's support services include instructional design, development and management, as well as hosting services, complete training, ongoing administration and 24/7/365 technical support for both students and faculty.

eCollege's staff is able to leverage its knowledge and ease the barriers of entry that many institutions experience when building their own online programs from the ground up. This allows eCollege's customers to concentrate on what they do best--offer curriculum, instruction and student services—and rely on eCollege to handle accessibility, scalability and availability of the eLearning program.

eCollege's Educational Partners include such institutions as National University; University of Massachusetts Amherst; University of Colorado; Drexel University; Montana State University-Billings; DeVry University, Inc.; Kentucky Virtual High School; and Microsoft Faculty Center. The company was founded in 1996 and is headquartered in Denver. For more information, visit www.eCollege.com.

REFERENCES

Budzynski, C.J. & Zabora, B.J. (2000). *eLearning: The Evolution of Education and Training* (Industry Analysis, Fall 2000). Baltimore, MD: Legg Mason, Inc.

Cattagni, A. & Westat, E.F. (2001). *Internet Access in U.S. Public Schools and Classrooms 1994-2000* (Report No. NCES 2001-071). Washington, DC: National Center for Education Statistics.

Consumer Electronics Association. (2001). *Technology and Education Survey*. Arlington, VA: Consumer Electronics Association, eBrain Market Research.

Lacy, K. (2001). *MSUB Online Program Review 1998-2001*. Unpublished manuscript, Montana State University–Billings.

Moe, M.T., Bailey, K., & Lau, R. (1999). *The Book of Knowledge: Investing in the Growing Education and Training Industry*. New York: Merrill Lynch & Co., Inc.

Shop.org (n.d.). *Statistics: US Internet Usage, Nielsen/Net Ratings*. Retrieved June 2001 from www.shop.org/learn/stats_usnet_general.html.

Tapscott, D. (1998). *Growing Up Digital: The Rise of the Net Generation*. New York: The McGraw Hill Companies.

Webber, S. & Boggs, R. (2001). *Higher Ed Technology Update: Networking, Distance Learning and Beyond* (Web Conference, January 31, 2001). Framingham, MA: IDC.

Appendix I

Online Survey Results

Mark Sheehan
Montana State University, USA

In summer 2001, the editors of this book conducted an online survey to gather opinions about what a portal is perceived to be in the context of higher education. Survey participants are listed in the Acknowledgments section of this book. The survey results were presented in a poster session at the EDUCAUSE 2001 conference and are summarized in Table 1 of the Introduction to this book. This appendix presents a more detailed analysis of the survey, including several items that were not included in the summary in the Introduction.

The survey was made up of 18 statements to which respondents were asked to react, and six questions that respondents were asked to answer.

Analysis of the survey results must be tempered by the fact that survey respondents were self-selecting. Most presumably had some interest in portals, and therefore likely had some familiarity with them. While survey respondents were in no sense hand-picked, they were at the same time not entirely randomly selected.

Statement 1.
To be considered a portal, a website must give me the option to customize it to look exactly the way I want it to.

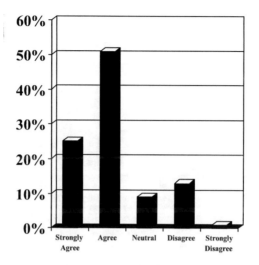

About 75% of respondents agree that a website needs to allow some customization by the user in order to be called a portal.

Statement 2.
To be considered a portal, a website must greet me by name when I access it (after my initial login).

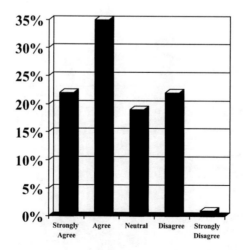

About 55% of respondents agree that a portal needs to recognize its users by name once the users have set their accounts up. A significant number (20%) disagree, though few disagree strongly.

Statement 3.

"Portal" is just another word for a website that supplies links to many other pages.

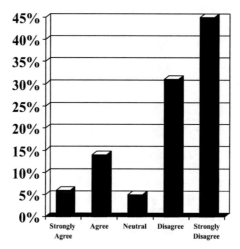

Over 75% of respondents reject the idea that the difference between a portal and a website is merely nomenclatural. Even so, another 20% appear skeptical that there is any real difference between a portal and a website.

Statement 4.

To be considered a portal, a website must accumulate information about my use of it AND customize accordingly its subsequent presentation of information to me.

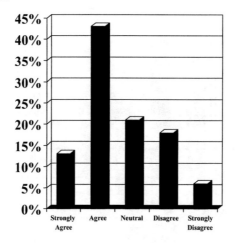

About 55% of respondents feel that a portal should track its members' usage and base changes in the way information is presented on the information gained thereby. Over a quarter of respondents disagree.

Statement 5.

To be considered a portal, a website must be "intelligent," showing me only information and links that I often use.

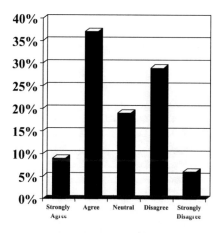

Opinions are mixed on this question. Over half of those expressing an opinion agree that a portal should monitor the member's use and limit what it displays to information and links often visited. Only slightly fewer appear to feel this is either unnecessary or inappropriate "behavior" on the part of the portal.

Statement 6.

To be considered a portal, a website must not display advertisements.

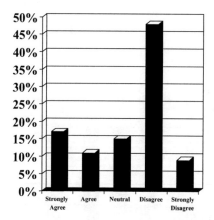

Well over half of respondents appear to feel that the incorporation of advertising does not disqualify a website from being a portal. Because so many current examples of portals *do* incorporate advertising, it would seem to be almost obvious that this is the case. Interestingly, over 25% of respondents appear to feel a website should *not* be considered a portal if it incorporates advertising!

Statement 7.

To be considered a portal, a website must be linked to a database of information about me, and as my characteristics in the database change, the website must present different information to me.

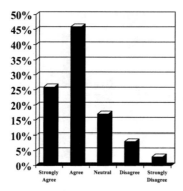

This statement is similar to statement number four, but varies from it in that the information about the member it refers to is not necessarily derived from the member's use of the portal, but could instead come from other sources, for example the university's personnel office. Responses to this statement are in fact very similar to those for statement four. Most respondents agree at some level that a portal needs to adapt the information it displays to a member to recorded information about the member.

Statement 8.

To be considered a portal, a website could be accessible from a personal digital assistant (e.g., a Palm Pilot).

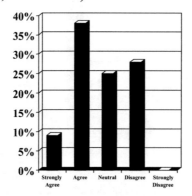

This statement was designed to examine the rigidity of the respondents' association of the word, portal, with the desktop computing environment. Almost half of the respondents felt a portal could be based on a smaller, mobile platform. Interestingly, though, over a quarter of respondents appear to feel that a portal is a desktop computer phenomenon. (See statement nine, below, for reinforcement of this analysis.)

Statement 9.
To be considered a portal, a website could be accessible from a Web-enabled mobile telephone.

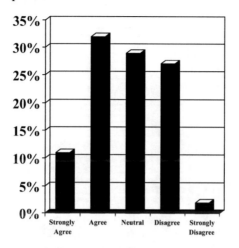

Opinions about the eligibility of mobile telephones as hosts for portals appear to be slightly more intense than those about PDAs, based on the larger "strong" responses for both agreement and disagreement. Overall, a slightly larger proportion of respondents (almost 30%) rejected the telephone as a portal host than rejected the PDA (25%).

Statement 10.
To be considered a portal, a website must function as an agent for me or as a personal assistant, helping me with my daily electronic communication needs.

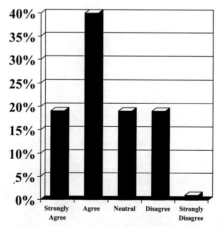

A majority of respondents expect a portal to exhibit some features of an "intelligent agent," in helping with daily communications needs. Fewer than 25% of respondents disagree.

Statement 11.

To be considered a portal, a website must require me to identify myself to it (i.e., authenticate with username and password or PIN) each time I access it.

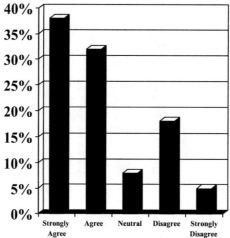

About 70% of respondents expect a portal to require its members to identify themselves each time they use it. Statements 12 and 13 follow up on this in order to sort out whether those who disagreed felt the portal should allow guest access or instead felt that some automatic authentication (a cookie stored on the member's computer, for example) could substitute for a manual login.

Statement 12.

To be considered a portal, a website must allow people to use it without logging in.

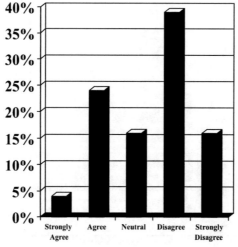

More than half of respondents felt that guest logins to a portal were not appropriate.

Statement 13.

To be considered a portal, a website must recognize me automatically, at least after my first login, and not require me to log in each time I connect using the same computer.

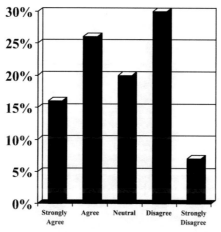

Opinion was almost evenly divided between those who felt a portal *must* in some way automatically identify and authenticate registered members (42%) and those who felt that was unnecessary (37%).

Question 14.

By what date in the future will most schools and companies replace their main websites with portals?

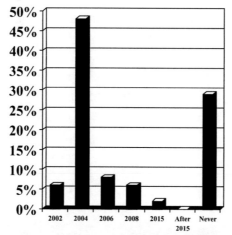

A clear majority of respondents felt that replacement of standard websites by portals was likely within the next 15 years. Significantly, though, over a quarter of respondents disagreed. A useful follow-up question would have been whether those who disagreed felt that portals are a transient phenomenon, or felt that even with a portal, an institution would still need a traditional website as well.

Question 15.
What is the status of the portal project at your school or company?

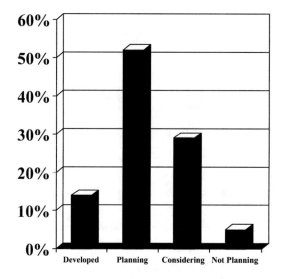

Only about 5% of respondents said their institutions were not considering, planning or operating a portal. More than half are either actively planning a portal or have brought one online.

Question 16.
What are some good examples of portals? (Fill in the blank.)
First place (8 of 41 citations):
- Yahoo.com

Tied for second place (2 of 41 citations each):
- Amazon.com
- Excite.com
- MSN.com
- My Netscape
- MyUBC
- MyYahoo
- UMN.edu
- UT Direct (utexas.edu)

Statement 17.
Amazon.com is a good example of a portal.

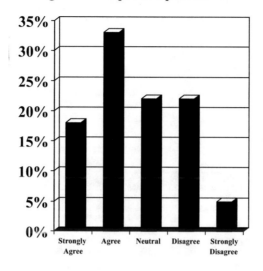

Just over a quarter of respondents rejected Amazon.com as an example of a portal.

Statement 18.
My UCLA is a good example of a portal.

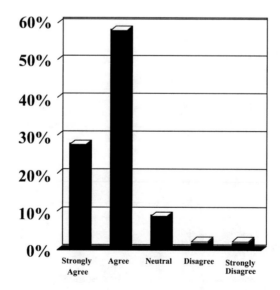

MyUCLA was overwhelmingly acknowledged as a good example of a portal.

Statement 19.
Excite.com is a good example of a portal.

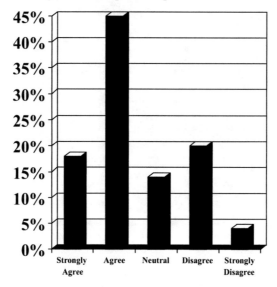

Nearly one-quarter of respondents rejected Excite.com as a good example of a portal. This is surprising because Excite.com was one of the first websites to refer to itself as an Internet portal, and as such should presumably have a certain right to that designation.

Statement 20.
My Yahoo! is a good example of a portal.

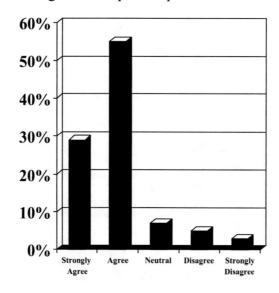

My Yahoo! was clearly acknowledged as a good example of a portal.

Statement 21.
CNN.com is a good example of a portal.

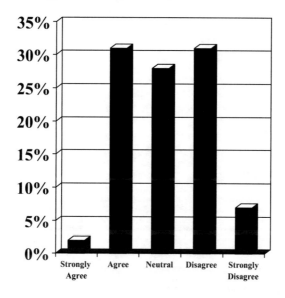

Opinion was almost evenly mixed about CNN's website. A small majority felt it did not deserve to be referred to as a portal, and relatively few (2%) respondents felt "strongly" that it should.

Question 22.
What are your top requirements for higher education portals?
Responses fell into five categories (of 239 total):
- Features and content provided (89)
- Customization of appearance (26)
- Personalization of content presented (35)
- Adaptivity to user's role (8)
- Other features (81)

In the "other features" category, the most frequent requirements mentioned were:
- Single sign-on
- Easy to use
- Secure
- Stable
- Flexible
- Few or no advertisements

Question 23.

What are your top requirements for commercial portals?
Responses fell into five categories (of 135 total):
- Features and content provided (26)
- Customization of appearance (14)
- Personalization of content presented (21)
- Adaptivity to user's role (3)
- Other features (73)

In the "other features" category, the requirements mentioned were essentially the same as for Question 18.

Question 24.

Do you have other comments?
- We are attempting to develop a "scholars' portal" that will provide a single entry point to a range of Web-based information services/collections.
- If everyone wants to offer a portal that will serve as the user's default starting page, which page will the user likely choose? There are just too many people trying to hijack users' starting pages so they can grab those eyeballs for ad-revenue-generation purposes. What users REALLY would like, I think, is a comprehensible list of starting links (à la Yahoo), plus a good search engine (à la Google).
- I think the idea is to have at your fingertips the links or buttons you need to get to all the information and applications you need for work and play. It includes access to information to do your job or be a student, and to engage in leisure activities. Thus it would have links to relevant discussions; to e-mail; to classes you are taking; to news of campus, city, state, national, world, etc. events; to chats if you do that sort of thing; to your student information, to corporate databases you need to access; to collaboration tools; to info about projects you are working on; to your calendar, etc.
- I also think the concept of a portal is still emerging. However, there is no question that organizing what you need to access in one place that allows you to get a bird's-eye view, find what you need easily, and drill down to information sources and applications has real value, no matter what you call it.
- We have looked at this issue for the last two years; in fact we started implementing a commercial product, "Campus Cruiser," and backed off it because the functionality of the built-in e-mail and calendaring systems were significantly less that we have in existing programs. We would really

like to have a portal that did not require that we change e-mail or calendaring software, but that those services could be integrated into our portal solution.
- Use uPortal—it has the best architecture.
- Very tough to find a commercial solution at this time. A very immature marketplace. Lots of smoke and mirrors.

Appendix II

Educational Portal White Paper

Ali Jafari
IUPUI, USA

ABOUT THIS WHITE PAPER

As mentioned in the Preface section of this book, in late 2000, the author developed and presented this white paper to the Indiana Higher Education Telecommunication System (IHTS). This white paper produced much conversation and received acceptance and support from a number of educational and government institutions in the State of Indiana. At the time of publishing this book, the IHETS has received funding to create a RFP and proof of concept as the initial stage of developing an Educational Portal for the State of Indiana. The new running name for this project has been changed to Indiana Learning Portal. More information about this project can be found at the IHETS website, http://www.ihets.org.

ABSTRACT

This white paper conceptualizes and discusses the design and application of a super web portal for state or nationwide educational applications. The term "Educational Portal" refers to a web gateway environment that allows users with varied educational interests to access educational resources and information. The Educational Portal provides a collaborative environment where educators can find peers who share educational knowledge and creativity. It is also a "profile-based" web environment portal that can be totally personalized according to each user's needs and interests, providing each user with specialized "MyPortal" functionality.

Many educational institutions are currently in the process of offering Web-based portals to their instructors and students. This might include services like course portal, library portal and campus portal. However, each of these portals offers services available within the institution only to the members of that institution. For instance, each portal can offer authenticated student and faculty access to courses, library materials and Student Information Systems in an institution. But there is no direct sharing of knowledge or resources between and among the members of various educational institutions.

It now appears necessary to offer a state or nationwide single front door portal gateway where any and all learners, regardless of institutional affiliation, can gain access to educational and training information. It will provide an opportunity for educators and trainers to share resources, and information and to collaborate on the development, evaluation and sharing of educational modules. For instance, the gateway will include a portal environment where a teacher can share the use of a personally developed e-Learning module with other teachers in the same field and class rank at other schools, or with students and others who wish to use it in their learning environment.

INTRODUCTION

There is a need for the design and implementation of an "Educational" Portal that offers three primary services within one portal environment. First, it will offer a place to find and acquire various educational resources, similar to services offered by amazon.com, but totally tuned toward teaching and learning materials. Secondly, the portal will promote exchange and trade of educational knowledge and creativity similar to functions offered by ebay.com, but optimized for exchange and trade of teaching, learning, and research knowledge and modules. The third component will feature a smart search engine that can find information according to your search parameters and your personal profiles, similar but more advanced than services offered by askjeeves.com. Placing these three primary services within one portal gateway will create a comprehensive educational gateway that can serve and link all educators at K-12 schools, colleges and universes. This super portal will complement other portal environments being built within colleges and universities called Campus Portals. Campus Portals offer some of these services only to the members of their particular institution. This white paper is intended to elaborate and justify the design and application of an educational portal to be a super web, serving the needs of teachers and learners independent of their school association.

WHAT IS AN EDUCATIONAL PORTAL?

As it is conceptualized in this paper, an Educational Portal is intended to provide a statewide or nationwide comprehensive front door access to educational resources and information. These resources include tools and services to facilitate the learning process as well as to complement teaching and learning. Learners can inquire about resources that are available to increase their knowledge and perhaps even collaborate with other learners with the same interests. The Educational Portal can, at the same time, provide a collaborative environment for the development, evaluation and sharing of educational modules. With this concept, for instance, a teacher can share a personally developed learning module such as a quiz, PowerPoint presentation, paper, streaming audio or video clip with peers in other schools while receiving feedback, peer review evaluation and usage log data.

An Educational Portal, in this paper, is conceptualized as a not-for-profit service offered to learners and teachers, supported by state, federal government or grant funding. A learner will come to the portal to find an educational resource. The learner will be given the opportunity to submit his or her personal profile so the portal might better serve him or her. If the learner is a teacher or faculty member, they can gain access to collaboration tools, course management tools and other "My Portal" type functions.

Teachers can register online using their school email address to validate their affiliation with an educational institution. Teachers can also register their students by uploading their class registry directly to the portal. The educational portal, upon verification of an email domain address, automatically sends a personal password to the user's email account. This automated registration process provides a dynamic and maintenance-free registration system to teachers and learners within state or national institutions. Learners not associated with an educational institution have limited access to information material.

An Educational Portal, in its full and complete concept, is a profile-based web environment. In its profile-based environment, the Educational Portal automatically places a user in a group category, based on the email address of an individual instructor. For instance, jsmith@mail.ips.k12.in.us places the user John Smith in the k-12 category of IPS (Indiana Public Schools) group. Furthermore, as part of the registration process, the user can provide additional personal information such as subject field, grade taught, research interests, conference interests, education level, gender, age, etc. The more demographic and professional information a user provides, the more personalized and filtered the information will be that the person receives.

The Educational Portal will also serve as an umbrella gateway to other existing portal services already offered by other agencies and institutions.

WHY CREATE AN EDUCATIONAL PORTAL?

- **Create a statewide gateway to online educational information.** The environment provides a gateway portal to educational information available form all educational institutions in the state (K-16, including Indiana Community Colleges). This would be a primary tool for learners to locate and register for credit and non-credit courses from learning institutions.
- **Create a statewide gateway to online educational resources.** The environment provides a gateway portal to statewide library resources for teachers and learners. For instance, this can include access to SAT practice tests for high school students, or links to online research tools and library resources for teachers and learners.
- **Provide peer reviewed collaborative environment.** Academics are actively encouraged or obligated to receive peer reviewed evaluations of their scholarly and creative works. This is usually mandated in the tenure and promotion process at many higher educational institutions. The Educational Portal provides the distribution, management and evaluation environment for peer review of educational, scholarly and creative works.
- **Share e-Learning modules.** Instructors can share the use of personally developed learning modules (hereafter referred to as e-Learning modules) with their peers in their schools, state, national or public level. E-Learning modules include Web-based resources in any electronic format, such as a file document, PowerPoint presentation, assessment quiz, streaming audio and video file, or any other multimedia format that can provide resources for online or distance learning courses.
- **Access to statewide e-Learning modules.** Instructors receive dynamic access to various e-Learning modules as they create or improve online courseware. An instructor, for instance, can find several e-Learning modules or assessment tools appropriate for his/her online course created and offered by other instructors in the same field and grade.
- **Receive automated educational news and information.** Teachers and learners receive automated news and information as it relates to their teaching, research and learning needs. The information can include new e-Learning modules developed by other instructors or those released by information providers, collaboration opportunities for research and course development, new online courses and degrees, etc.
- **Create statewide metadata of e-Learning resources.** As more resources are catalogued in the environment, the Educational Portal creates a comprehensive database containing information about e-Learning modules, courses and other educational applications.

- **Create a statewide teacher and learner database.** As more teachers and learners sign on to the Educational Portal, a comprehensive list of teachers and learners interested in e-Learning will be maintained in a database.
- **Provide incentives to engage teachers in development and sharing of educational modules.** Providing real-time usage data to instructors who post their e-Learning modules or courses establishes an incentive to create and share more modules, tools and skills.
- **Provide centralized access to comprehensive course management tools.** The Educational Portal provides comprehensive course management and teaching and learning tools for those whose institutions are not yet offering course management software. Every instructor receives full access to course management, Web authoring and assessment tools to create distance learning courses or to complement traditional lecture-based courses.

WHAT ARE THE INCENTIVES FOR USING AN EDUCATIONAL PORTAL?

- As an instructor, I will have access to use a wider selection of e-Learning modules and resources developed by other instructors in my field and grade level.
- My e-Learning modules will be used by my peers and other students (not just my students), and I will know the exact usage data through MyPortal. Knowing the usage data is beneficial for my tenure and promotion process.
- I can receive external assessment of my creative and scholarly work, which is also valuable information needed for my tenure and promotion process.
- More colleagues, besides those in my school, will learn about my work, expertise and interests. As more people know me and know my work, I will find greater career opportunities and more off-site consulting available, should I desire it.

WHY WOULD TEACHERS USE THE EDUCATIONAL PORTAL?

- Many schools and universities are in the process of offering campus portals. By default, the campus portal is meant to serve the purposes of students and teachers within a campus or university system. Campus portals are not designed to build collaboration outside of the individual institution or provide a mechanism to share knowledge and resources among educational groups outside of a school or university environment. For instance, there might be

only five biology teachers accessing their campus portal, but through the Educational Portal, the five biology teachers will increase their ability to collaborate with a much larger group of biology teachers and share resources on the state or national level.
- Schools may not be able to provide their teachers and students with all the teaching and learning resources they may need. On a state level, operation of the Educational Portal can provide statewide resources both in terms of available courses from all educational institutions (the Virtual University), technology software (course management), library resources (online information resources), technology services (file servers and Web servers) and collaborative environments (focus groups) to every educator, above and beyond the boundaries of an individual school.
- The Educational Portal provides "MyPortal" access to educational resources and information on the state and national level, with one single front door gateway to educational resources, information and collaboration.

WHO SHOULD OFFER AND MAINTAIN AN EDUCATIONAL PORTAL?

Ideally, an organizational entity with statewide responsibility should fund and support an Educational Portal. The web environment is meant to be self-maintainable, allowing each instructor to archive self-owned e-Learning modules and set the access rights. With this concept, the management, access and control are set by each individual member and is programmatically maintained by the system software. The system software automatically maintains membership accounts as well. If a user loses his/her school job, it is assumed that his/her email will be cancelled and therefore the system will automatically deny or limit the access to the user.

WHO ARE THE PRIMARY USERS OF AN EDUCATIONAL PORTAL?

The primary members of the Educational Portal are learners in Indiana and educators associated with a K-12 school, college or university. Learners represent all walks of life, seeking credit or non-credit coursework, undergraduate or graduate degrees, or just ways to otherwise improve their skills. Educators include instructors, researchers, librarians and administrators. Another user group includes students whose portal association and membership is initiated through a member teacher. Parents whose membership can be initiated through their children's membership can also access the portal. The final group includes adult learners and public community members who will have access to various public resources and information available through the main portal interface.

HOW DOES ONE BECOME A MEMBER?

Instructors and students associated with educational institutions can register and become a member by visiting the Educational Portal homepage and entering their school email account. Users seeking to be a member of the portal will be asked to enter basic information such as grade taught, fields of instruction and research, interest groups, etc. Upon the completion of the form, the user's password will be automatically emailed to the user. With this concept only those associated with affiliated schools will receive automatic registration. Once a teacher creates a course or a collaboration group, she or he can invite members, such as students, into the class. The learners or group members will then automatically receive email confirmation about their membership with a class or group. Their password, along with other information and instructions, will be sent via email to users, and their username, as always, will be the full domain email address of the individual. Similarly, parents can receive membership access to the portal if their child has a membership account to the Educational Portal. Students or parents not having a school email account will obtain a username and password authentication information through the teacher or the group leader.

With this system design model, parents can receive membership accounts for the Educational Portal if their child has a membership account. The children (students) get a membership account if they are class members of a course and if a teacher has entered their name in the class roster. The teachers get a membership account if they are with an affiliated school and if they use their school email address to register. In this model, it is assumed that teachers have email accounts from their school, and the portal system software maintains the list of all the affiliated schools and their domain email addresses. Beside teachers, students and parents, other interest groups, such as adult learners, can receive access to the Educational Portal through the public site or registration form.

WHAT ARE THE PRIMARY RESOURCES OFFERED WITHIN AN EDUCATIONAL PORTAL?

The primary resources offered within an Educational Portal include:
I. Portal Services:
    A. MyPortal (a totally personalized, customizable and dynamic website capable of offering customized, channeled and pushed information)

B. Personal disk space (in the form of a file, Web and streaming media server) for achieving and sharing e-Learning modules
C. A file sharing environment for peer evaluation of scholarly and creative works
D. Tools for communication and collaboration (message boards, group list servers, email, chat rooms, etc.)
E. Course management software
F. Group collaboration software
G. Assessment, grade book, attendance, and other teaching and learning tools

II. Teaching and Learning Resources:
A. E-Learning modules created by instructors
B. Distance learning courses and online professional development content
C. Licensed third-party teaching and learning libraries, and online resources
D. News and information

III. Other Information:
A. Educational news channeled or pushed by the system software or individual members
B. Catalog of courseware and professional development modules
C. Metadata of e-Learning modules

WHAT ARE THE FUNCTIONAL AND TECHNICAL REQUIREMENTS OF AN EDUCATIONAL PORTAL?

- Ability of average user to master the environment without training or technical assistance
- Ability for automatic primary member (instructors) to register using a school's email account
- Ability for primary members to invite new members (students)
- Ability to archive (store) e-Learning modules (electronic files) via the Web interface in various files, streaming media and multimedia formats
- Ability to set the sharing rights for each uploaded electronic file to allow sharing with members within the same school, the state, the nation or to allow public access, as well as "no-sharing" rights
- Ability to input metadata information on type, subject, grade, etc. as specified by the Indiana Educational Standard and EDUCAUSE IMS specifications

- Ability for each user to view log data information on who has visited his/her archived files
- Ability for each user to view the log data on showing the total number of uses of each e-Learning module and the total number of visits to MyPortal.
- Ability to review and evaluate the work of other members in the environment
- Ability to set filters to automatically receive channeled news information according to a member's personal interests, fields, grade taught, etc.
- Ability to post news and classified ads entered by field, interest, grade taught, etc.
- Ability to set share and view rights for various resources within MyPortal, including personal information, bookmarks, news, files, etc.
- Provide an extensive set of communication, collaboration and course management tools
- Provide necessary controls for customization and personalization of the MyPortal interface
- Provide advanced utility tools, like the Instant Messenger function, for real-time locating certain online members or groups
- Ability to program a series of intelligent agents to further automate and personalize the use of the environment
- Ability to program a series of intelligent agents with autonomy for making decisions about portal management
- Ability for future integration with school database systems for assistance in authentication, personal information data, institutional data, etc.
- Ability to interface with school email systems for email retrieval and filing

HOW DOES THE USER INTERFACE LOOK?

The main page of the Educational Portal would have an easy to remember homepage address, a .org domain name. The main Educational Portal homepage, here referred to as the primary interface, includes the following main section and links:

- **Logon field.** A logon field to enter username and password is immediately followed with text links for "Problems Signing In?" and "Forgot Your Password?"
- **New Educator Accounts.** A text link for "New Educator Sign Up" with a popup page for new online membership applications. Educators are required to use their school email address to obtain their portal membership account. Email addresses without a pre-approved school domain name will automatically be rejected.

- **New Student and Parent Accounts.** A text link for "New Student and Parent Sign Up" that provides information on how to obtain membership portal accounts for new students and parents, and the qualification guidelines with appropriate online application forms.
- **About Educational Portal.** A link providing information about the Educational portal project, purpose, ownership, etc. Other links can be provided to include information about who should use the portal, available resources and services, privacy and copyright information, etc.
- **Intelligent Search Engine.** A search field box linked to an intelligent search engine to access both internal and external teaching and learning resources.
- **Public Resources and Services.** A link(s) providing access to various resources and services available to public users. No authentication (login) is necessary to access these resources.
- **Public Information News and Calendar.** A frame or a portion of the main page to include public news information. No authentication is necessary to obtain the news on the main page.
- **Other Links and Fields.** Additional links and fields as identified by the Educational Portal stakeholder as necessary for public users.

MYPORTAL PAGE

The concept for the front portal page (the primary page), described earlier, is to provide information and resources for those who do not have membership accounts with the Educational Portal and for members to sign into the system. Members must sign into the system in order to get access to the group and personal resources.

Once a member has signed in, she or he will receive a personal portal page, here referred to as a "MyPortal" page. The MyPortal page (see Figure 1) offers different features, resources, services and appearances [formats] for instructors (educators), students (learners) and parents. An owner of a MyPortal page can customize and personalize his/her MyPortal and receive various controls and monitoring data as listed below. Figure 1 shows an example of a teacher's MyPortal page.

User Verification and Access Data

Upon a successful login, the user will be greeted by name, and his or her position and institutional affiliation, followed by the most recent login activities and statistical information about MyPortal hits or visitors. See Figure 2.

Figure 1. Personal Portal Page Example (MyPortal) of an Instructor

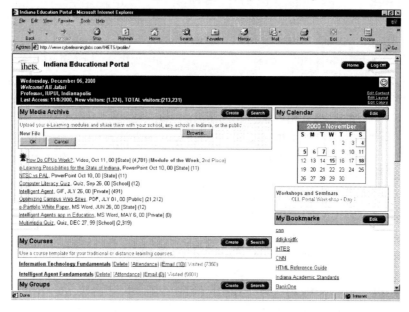

Figure 2.

MyPortal Customization

The user will be able to customize the appearance and organization of services and contents by clicking on various edit links. See the right section of Figure 2.

Personal Information

The user will be able to edit and further complete personal information such as contact and demographical information, school, education, field of study, research interests, teaching assignment, conference interests, etc. The more personal and professional information a user provides, the more dynamic and filtered the information will be presented by MyPortal. Figure 3 represents a simplified version of this form.

Intelligent Agents

The Educational Portal is conceptualized as a "smart" portal environment offering a series of intelligent agents. A user can program his personal agents to

Figure 3. A Simplified Version of a Personal Information Form

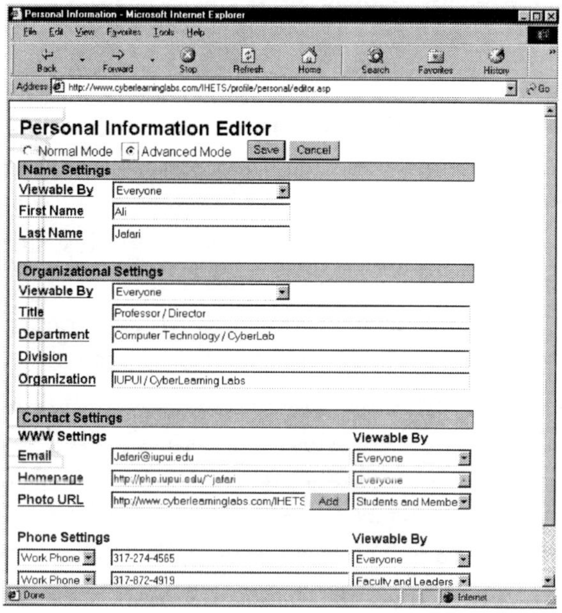

perform certain tasks and have agents make certain decisions based on specified criteria. Intelligent Agents will be proposed and designed for various tasks and uses within the Educational Portal environment. See Figure 4 for an example of agents listed for course management applications.

Figure 4. A Series of Intelligent Agents as Appropriate for Course Management Environment

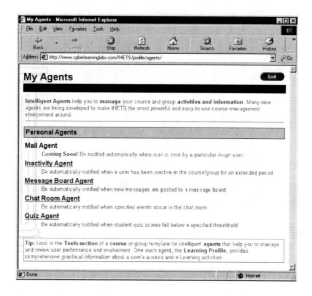

The following list includes resources and services within a MyPortal page. The following items are mostly conceptualized for an instructor (educator) MyPortal.

My Media Archive

The My Media Archive provides a file and Web server service and necessary management tools to archive, set access rights and manage electronic files at a central server(s). These files can be formatted using Word, PowerPoint, text, HTML, graphics, quiz, streaming audio and video files, and other multimedia formats. The user clicks on the Browse icon (see Figure 5), selects a file from the local disc, completes a metadata information form, sets access rights and uploads the file. The user will select access right levels from categories including private, school, state, national and public. Other levels can also be included. By selecting a "state" access level, for instance, the environment will allow access to the file to anyone within the associated schools on the state level. While the "school" access level will only provide file access to instructors within the user's school, "private" access level does not provide access to anyone except the owner of the file.

As shown in Figure 5, My Media Archive provides usage data for every file in the system. For instance, if a file (e-Learning module) was included in a different teacher's course other than the owner of the file, and was accessed by 25 students, the access log data will be incremented by 25 and the owner of the file will know that 25 new people have used his/her e-Learning module.

The Educational Portal can offer promotional prizes like automatic identification and notification of the first, second and third place of the most used e-Learning modules and notify the owner. The system can even automatically print a certificate and mail it to the owner for his/her dossier filing. Furthermore, the system, upon previous arrangement with .com companies, can automatically send, for instance, a $100 gift certificate from Amazon.com to the first place weekly winner of the most used e-Learning module, $50 for the second, and so forth. More possibilities can

Figure 5. The My Media Archive Service

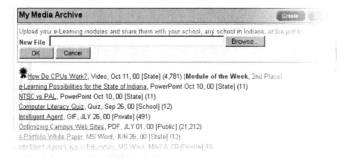

be speculated upon to provide incentives for teachers and scholars to create, share and promote the use of their creative and scholarly works.

My Courses

The Educational Portal can provide a comprehensive course management tool. The course management tool can be useful for schools where this service is not available or where there may exist a less sophisticated and more difficult to use course management system.

The MyPortal page provides real-time log data showing how many times a course is being used with direct links to a grade book, attendance page (useful for K-12 courses) and course email. Each informational text is hyperlink to the linked application. See Figure 6.

The course management software provides various authoring and management tools for putting course contents online. The course management software can

Figure 6. My Courses Section Provides Direct Link to a Course Template in Addition to Statistical Information Regarding the Number of Hits and Direct Link to Course Mail, Attendance, Etc.

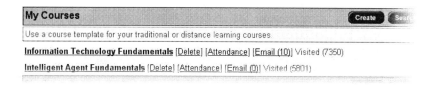

Figure 7. Course Template

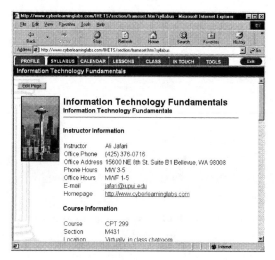

Figure 8. Educational Portal Offers Group Collaboration Management Tool Similar to the Course Management Tool

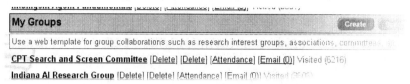

Figure 9. A Group Collaboration Template

be used for distance learning courses or as a Web template to complement a traditional classroom-based course or training curriculum.

If a user wishes to utilize a third-party course management system, not integrated with the Educational Portal software, the My Course section may be used to provide hyperlink access to the third-party course management server.

Figure 7 shows a course management template with categories appropriate for higher education courses.

My Groups

The Educational Portal provides a template environment for group communication and collaborative activities. Examples include a group of researchers collaborating on a research subject sharing their interest in a research field.

The MyPortal page provides real-time log data displaying how many times a group template is visited with other direct link as shown in Figure 8.

Similar to the course management tool, the collaboration group provides authoring and assessment tools for group collaboration. Figure 9 shows the group template page.

Figure 10. Educational Portal Provides a Collaboration Tool for Peer Review of Scholarly and Creative Works

My Reviews
Receive peer reviews of your scholarly or production works.
Good Quiz!, Jana Hickey, Pike, Oct 11, 00
Good paper Ali..., John Smith, IUB, Oct 30, 00
My students loved your paper, Bill Chism, IUK, Oct 31, 00
Highly recommend your AI paper, John Kern, IPS, Nov 09, 00

Figure 11. Educational Portal Provides a Hyperlink to Various Email Accounts Used by a Member

My Email
Create shortcuts to your varies school or commercial email accounts
My E&T Email (Exchange)
My CyberLearning Labs Email
My Yahoo Email
My IUPUI Email (Exchange Account)
Information Technology Fundamentals (16)

My Reviews

An important feature offered by the Educational Portal is a software tool providing a complete mechanism for peer review of scholarly and creative work activities of educators. Once an educator achieves his/her creative or scholarly work on the Educational Portal, the work could be used, evaluated, reviewed and commended by other instructors. The evaluation notes automatically appear on the MyPortal page as shown in Figure 10.

My Emails

Many people use two or more email accounts for academic and work-related electronic communications. The MyPortal page provides hyperlinks to various personal email accounts as desired and edited by a user. See Figure 11.

Figure 12. Educational Portal Provides Dynamic News Posting in MyPortal Page

My News
Receive filtered news or post items that relate to your field or research interests.
IU extends partnership with Microsoft
The Office of the Vice President for Information Technology has announced an extension of IU's highly successful enterprise license agreement with Microsoft, continuing the groundbreaking agreement until June 2003. The extension of this agreement demonstrates the University's commitment to improving its technology infrastructure and to providing students, faculty, and staff with the latest and most popular technology.

Figure 13. Educational Portal Provides Personal Calendar to Each Member

My News

The My News section of the Educational Portal can be setup to automatically display a channeled or pushed listing of news as desired by a member or as it relates to the member's field and grade level. The news listing dynamically appears on the MyPortal page of a member and the system automatically updates the listing as current and more related news is posted on the environment. See Figure 12.

My Calendar

Each member receives a personal calendar with his/her Educational Portal account. The Calendar can be used for personal or professional applications. The MyPortal page can automatically show the daily or weekly activities on the MyPortal page. See Figure 13.

Figure 14. Educational Portal Provides Virtual Bookmark on MyPortal Page

My Bookmarks

The Educational Portal offers a personal virtual bookmark listing as part of the features available on the MyPortal page. The Educational Portal may dynamically add additional bookmarks into one's personal bookmark as they are related to the field, grade level and research interest of a member. Educational Portal members can edit and add additional bookmarks as needed. See Figure 14.

Figure 15. Educational Portal Research and Development

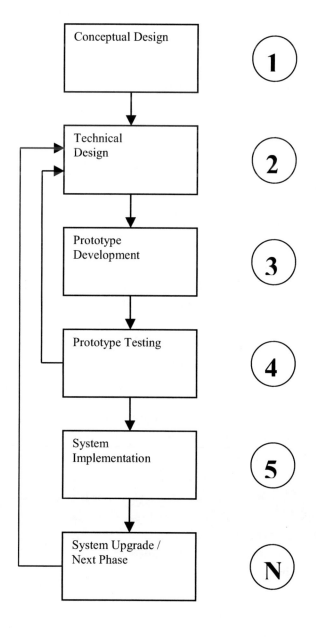

OTHER TOOLS AND SERVICES

The Educational Portal can offer additional tools and services as needed.

What Should Be the Design and Development Process?

As indicated in Figure 15, the research, development and implementation process of the Educational Portal consists of five steps. They include:

1. Conceptual Design

The conceptual design phase includes the development of a white paper to define the Educational Portal functional and technical requirements from the conceptual perspective. After the development of the white paper, it will be necessary to discuss and refine the proposed conceptual solutions in brainstorming sessions with various stakeholders. The stakeholders are leaders and subject matter experts from state institutions who will lead the implementation of the Educational Portal project. Other groups included in the brainstorming sessions are educational and statewide service provider organizations or institutions such as statewide IT service and solution providers, K-12, colleges and higher education institutions.

Tasks include:
- Identify various stakeholders and institutions to form a working group.
- Review and finalize various services and resources currently being offered through the Educational Portal environment.
- Review and finalize the design of user interface and navigational procedure.
- Identify and address various rules and regulations in terms of ownership and copyright, member responsibilities and obligations, etc.
- Identify operational practices for dividing responsibilities among various educational and organizational institutions.

2. Technical Design

Refinement and approvals or conceptual solutions identified in the white paper should be used to develop the technical design document for the design of the Educational Portal environment. The technical design phase addresses questions about the technology platform, software architecture, database design, database business rules definition, integration requirement and the like.

Tasks include:
- Identify technical requirements and specifications for the Educational Portal in terms of capacity, lead, maintenance, etc.
- Develop hardware and software specifications.

- Identify new tools and features not included in an off-the-shelf software for in-house or third party development.
- Identify and address all integration design and security issues.

3. Prototype Development

The prototype development phase includes the construction of the Educational Portal for a small-scale platform. The main purpose of the prototype development phase is to simulate the hardware and software systems to verify and measure the expected requirements before constructing the final enterprise system. The prototype system may not offer the "horsepower" to support a large number of users, but it should provide the main functionality the final system is expected to deliver.

Tasks include:
- Specify and build prototype hardware and software environment.
- Build prototype system.

4. Prototype Testing

The Prototype testing phase includes both technical and functional usability testing of the prototyped system. A selected sample of users, representing users of every major group, should be selected to use and evaluate the prototyped environment. The usability testers will provide important feedback for debugging and further refinement of the system. The most important functional requirement of the system is "ease of use." The prototype-testing phase should measure the ease-of-use factor and identify alternative measures to improve the usability of the system for every user group.

Tasks include:
- Identify a methodology for usability testing.
- Select usability testers from various group users.
- Conduct usability testing and measure the variables.
- Document system, software, and user interface bugs and usability issues.

5. System Implementation

The last phase of the project includes development of the final hardware and software infrastructure and the establishment of support resources. This phase includes the identification of appropriate support groups of staff with defined responsibilities for various operational and maintenance practices.

Tasks include:
- Build final hardware and software system.
- Establish various support groups with definite responsibilities.
- Define the management and team infrastructure.

6. System Upgrade/Next Phase

It should be assumed that the Educational Portal system will require upgrades and refinements, as new technology emerges and new applications and services are requested. The operational budget should include line support staff for system upgrades and the development of new services, tools and applications.
- Identify research and development resources.
- Identify and secure appropriate budget.

About the Authors

Ali Jafari (jafari@iupui) has worked in the fields of Information Technology and Multimedia since 1985 as a software designer, system engineer, technology architect, professor and researcher. He has worked in various engineering, administrative and academic positions at the Indiana University Bloomington and IUPUI campuses. He is currently Director of the IUPUI CyberLab and Professor of Computer Technology in the School of Engineering and Technology at IUPUI. Dr. Jafari has presented at more than 100 national and international conferences and published in professional and scholarly journals on a variety of subjects in information technology as it relates to teaching and learning. Dr. Jafari's research interests include interface design, agent-based learning environments, intelligent user interfaces and most recently Internet portals. Dr. Jafari has initiated, directed and co-developed several major research and development projects including the Interactive Multimedia Distribution System (1997), the Oncourse enterprise course management system (1999) and the Angel e-Learning Portal (2000). Professor Jafari's new research and development project is electronic portfolios, a new teaching and learning portal environment to be built by a consortium of higher education institutions. For this purpose he has initiated and established the ePort Consortium (ePortConsortium.org).

Mark Sheehan (sheehan@montana.edu) is Executive Director for Information Services and Chief Information Officer for Montana State University. In recent years he has presented an introduction to portal technologies to audiences at EDUCAUSE 2000, the 2000 and 2001 User Services Conferences of the Special Interest Group for University and College Computer Services (SIGUCCS) of the Association for Computing Machinery, and the 2001 annual meeting of the Northwest Academic Computing Consortium. Dr. Sheehan writes frequently for such publications as *EDUCAUSE Review*, *EDUCAUSE Quarterly* and *ONLINE*. In 1995 he won the UMI Excellence in Writing Award for his article "Pulling the Internet Together with Mosaic," published in *ONLINE*.

* * *

Stephen Ast serves as eCollege's Director of Campus Services. Mr. Ast joined eCollege in 1996, and oversees the initial implementation of online campuses. His areas of expertise include training school administration, faculty and staff; working with the school's designated appointees to help work through policy issues, site content and layout of the online campus; and developing custom work for individual institutions including reports, online registration and payment, and back-office integration. Mr. Ast was recently involved in building the Montana State University Billings Online CampusPortal^{SM}. He earned his bachelor's of science degree in Human Resources Management at Ithaca College, and has presented at the past three Center for Internet Technology in Education conferences.

Robert Aucoin, MA, is Director of Distance Delivery in the Faculty of Rehabilitation Medicine at the University of Alberta, Edmonton, Alberta, Canada.

Anne Yandell Bishop joined the Information Systems Department at Wake Forest University in 1981 and led the implementation of many of its business software systems. After the introduction of ubiquitous computing on the Wake Forest campus in 1996, she originated the idea for the university's portal and managed its development, launch and ongoing enhancements. She authored a chapter entitled "Extending Computer Usage to Administrative Areas" for *Electronically Enhanced Education*, edited by Dr. David G. Brown and published by the Wake Forest University Press-Scientific Division in 1999. She has conducted a number of workshops and made numerous presentations on portals and Web-based class registration. In 2001 she became Director of Research and Development in Information Systems and turned her attention to emerging technologies in academic areas. Ms. Bishop is currently conducting projects to develop software for and assess the value of hand-held computers and other technologies in teaching and learning. She holds BA and MA degrees in Mathematics from the University of North Carolina at Greensboro and an MBA degree from Wake Forest.

Katy Campbell, PhD, is Acting Director of Academic Technologies for Learning and Associate Dean in the Faculty of Extension at the University of Alberta, Edmonton, Alberta, Canada.

Robert Duffner is Director of Product Marketing at BEA Systems, USA. He is the lead evangelist for the BEA WebLogic Portal and for initiatives in e-business infrastructure. Mr. Duffner is responsible for product marketing communications and sales channel productivity for the division's products. Prior to joining BEA, Mr. Duffner was Director of Product Marketing for Vignette Corporation. During his two years at Vignette, he helped build the product marketing organization and successfully executed Vignette's first global, multi-product launch. Prior to Vignette, he was Director of Product Marketing at the Vantive Corporation and Vice President of Strategic Marketing at Pangaea Software. Mr. Duffner holds a bachelor's degree in Biochemistry from the University of Maryland.

Stephen C. Ehrmann, PhD, is Director of the Flashlight Program and a founder and Vice President of The Teaching, Learning and Technology Group. For over 25 years, he has been helping educators improve teaching and learning. Since 1993, he has directed the Flashlight Program, which helps educators evaluate and improve their own uses of technology, on and off campus. Flashlight may be best known for its award-winning tools for developing evaluative studies. Dr. Ehrmann is also well known in the field of distance education, dating back to his years of funding innovative research and materials in this field when he served as a Program Officer with Annenberg/CPB (1985-96). Before that he was a Program Officer with The Fund for the Improvement of Postsecondary Education (FIPSE) and Director of Educational Research and Assistance at The Evergreen State College.

David L. Eisler has served as Provost of Weber State University, USA, since 1996. At Weber he has been directly involved with the creation of WSU Online, new curricular efforts and positive student enrollment patterns. Dr. Eisler serves on the Utah Education Network Steering Committee, and the Utah System of Higher Education Academic and Applied Technology Committee. Previously he was Dean of the Eastern New Mexico University College of Fine Arts and Assistant Dean of the Troy State University School of Fine Arts. He works actively with Teaching, Learning and Technology Roundtables, and writes and presents regularly on assessment issues, faculty/staff technology support, portals and the application of technology to solve teaching/learning problems. Dr. Eisler was the seminar leader for "Provosts on Portals," a Web-based learning experience which involved over 120 chief academic officers from the AASCU institutions.

Christopher Etesse, Director of Technology and Commerce, Blackboard, Inc., USA, brings a strong technical background to the development of Blackboard's technical strategy, including experience in both the academic and private sectors. Mr. Etesse's understanding of the education market also comes from direct experience as an instructor as well as a designer of enterprise institutional and course management systems. Mr. Etesse has taught C++ programming to undergraduates, developed Web-based education products, as well as helped launch enterprise-level educational suites. He has been heavily involved in e-Education from the earliest days of the Internet—from having one of the first course sites for students to retrieve course materials and check grades in the beginning of 1996, to architecting the first Course Management System for International Thomson Publishing in late 1997. He joined Blackboard in late 1998 and had helped nurture numerous technical strategies and initiatives within the corporation--from the initial technical service deployments, enterprise integrations, direct leadership in development of Blackboard 5 and the Blackboard Platform Builder, as well as principal conducting technical due diligence in mergers and acquisitions. Mr. Etesse received his Master's of Science in Computer Science from the University of Kentucky.

James P. Frazee is a doctoral student studying educational technology in a joint program between San Diego State University and the University of San Diego, USA. Formerly the Director of Information Technology for the Sweetwater Union High School District, the largest secondary school district in California in 2000, Mr. Frazee is now the Associate Director of Instructional Technology Services at San Diego State University. Teaching is a passion of his, and for several semesters he taught the "Technologies for Teaching" course at SDSU. He has presented widely on the subjects of obtaining, managing and leveraging large federal educational technology grants; designing faculty professional development programs; and using technology to improve communications, extend participatory leadership and push curriculum reform. His current research focuses on the use of wireless technologies, particularly hand-held computers, to facilitate active learning and formative student evaluation of faculty in higher education. He is a member of the Directors of Educational Technology / California Higher Education (DET/CHE), the College Consortium of University Media Centers (CCUMC) and EDUCAUSE. You can reach him at jfrazee@mail.sdsu.edu.

Rebecca Vaughan Frazee is a consultant to corporate and nonprofit organizations. Specializing in performance analysis and e-learning, she has managed projects for many global and Fortune 50 companies, the IRS and the Corporation for National Service. Her work centers on the development of training and education professionals, helping them expand their focus beyond training to more systemic performance solutions and tackling questions such as: "What skills are necessary to shift from training to performance? How can technology be leveraged to support these professionals and share central resources across geographic and functional boundaries?" Ms. Frazee has conducted workshops, presentations and university lectures on project management, data analysis and performance improvement. She is currently pursuing a doctorate in Educational Technology at San Diego State University and the University of San Diego. You can find her article on technology adoption in the *2002 ASTD E-Learning Handbook,* and you may reach her at rebvaughan@att.net.

Cassandra Gerfen, Product Director of Campus Solutions℠, oversees the vision and direction of the product line based on internal and external feedback, the needs of the market, analysis of features and the company's overall goals. Ms. Gerfen started working for eCollege, USA, in 1997 where she served as Marketing Communications Coordinator for one year and Director of Client Services for two years. Ms. Gerfen received her bachelor's of science degree in human resources management from Colorado State University, and her MBA from Creighton University.

William H. Graves is Founder of Eduprise, a CollegisEduprise Company, and Co-Chairman of the Board of CollegisEduprise, Inc. His perspective derives from more than 30 years of experience in higher education and from the trust earned by CollegisEduprise as a product-neutral partner in planning, implementing, managing and evaluating Internet-related services to meet today's higher education challenges cost effectively. He is a recognized for his leadership on the use of the Internet in teaching and learning, and in the services that support the educational process. Dr. Graves has given hundreds of presentations, advised hundreds of institutions and published more than 60 articles on technology-in-education themes. He is a member of the board of directors of CollegisEduprise, EDUCAUSE and the Instructional Management Systems Global Learning Consortium. He was a cofounder of EDUCAUSE's National Learning Infrastructure Initiative and still chairs the NLII planning committee. He also was one of the founders of the University Corporation for Advanced Internet Development and its Internet2 project. He earned his doctorate at Indiana University. He then served on the faculty of the University of North Carolina at Chapel Hill, and also as Dean for General Education, Interim Vice Chancellor for Academic Affairs, Senior Information Technology Officer, and founder and director of the Institute for Academic Technology, a partnership with IBM. He became Professor Emeritus of Mathematics at UNC-Chapel Hill upon leaving there with his institute colleagues to create Eduprise.

Kirsten Hale is Assistant Director of Web Integration Services at Eduprise. She is responsible for managing the development of integration solutions between course management systems, student information systems, e-commerce systems, library systems, portals and other vertical sources of data. She has more than five years of experience in managing technical projects, gathering requirements from clients and evaluating products in light of these requirements. She also has experience in the management of software development and testing and in the quality assurance of course management systems. Mr. Hale holds an MS in Technical Communication from North Carolina State University, where she studied the differences between online and face-to-face instruction as it relates to student performance and student satisfaction. She is currently working on her doctoral degree in Information Science at the University of North Carolina at Chapel Hill, where she is focusing on the development and maintenance of online communities.

David Sharpe is Director of Instructional Technology Services at San Diego State University, USA, where he has management responsibility for the operation of this 24-person unit. Prior to joining ITS in 1983, he taught instructional design and the production on instructional materials in the Educational Technology Department at SDSU for 6 years. He has consulted with both corporations and educational institutions concerning the design and use of instructional materials. His BA degree is in Telecommunications and Film, MA in Educational Technology and EdD in Instructional Systems Technology. He served as chair of the Ad Hoc Portal Committee that was established to provide recommendations to the Associate Vice President for Academic Affairs concerning the development of a campus-wide portal.

James Thomas received his Bachelor's of Science degree in Business Administration and his Masters of Public Affairs in Public Administration from Indiana University. He has more than 14 years of experience in developing and managing enterprise application projects at Indiana University, USA. Mr. Thomas is Manager of the Systems Integration Team (SIT) in University Information Systems (UIS) at Indiana University. SIT is responsible for strategic analysis and ongoing support to UIS in the areas of enterprise application integration, enterprise systems development environments, component-based development methodology, enterprise application architecture and user-centered design. Part of this role involves managing the enterprise application portal project, OneStart, which will provide a service delivery framework allowing universal access to online university services via a unified and personalizable front end. Portal services are built upon an infrastructure of reusable components with published interfaces called the Enterprise Development Environment or EDEN. EDEN is made up of components such as an integrated workflow engine, a sticky authentication service and standardized business rules. EDEN provides agility, scalability and extensibility, allowing the number of services to grow and adjust in the fast-paced world of information technology.

Jameson Watkins is Assistant Director for Internet Development at the University of Kansas Medical Center, USA. He also is on the national faculty of the School of Library and Information Management, where he teaches graduate courses in information architecture, Web design and TCP/IP networking. He is the Educause constituent group leader for Web portals, and participates in local and state Web initiatives like the Kansas Digital Library. He holds a Bachelor's of Science in English and a Master's in Library Science from Emporia State University.

Index

Symbols

2EE Connector Architecture (J2EE CA) 213
80/20 Rule 15

A

academic freedom 167
access control 194
accessibility 169, 255
accountability 164
action plan for your portal project 31
Active Server Pages 209
activity-based costing 34
ADA compliance 14
adaptive 124
admission 241
adult learners 240
advertising 3
affiliation 41
affordable usability 24
alumni 195
announcements 192, 247
application programming interface (API) 60, 246
application service providers 244
application vendors 208
artificial intelligence 90
asynchronous 167
authentication time 20
autonomy 90, 93
availability 12, 14
awareness 146

B

baseline studies 32
BEA Systems 208
blackboard 43, 131
bolt-on 38
branding 247
brochure Web sites 238
browser compatibility 21
build or buy 110
business portal 206

C

cache 249
CalStateTEACH 42
campus communities 105
campus homepage 7
Campus Pipeline 43
campus portal 9, 165
CampusSolutionsSM 242
career counseling 241
case study 133
cell (mobile) phones 99
channels 2, 53, 80, 245
classrooms 166
COHERE 176
collaboration 167, 242
collaborative learning 151
College of the Holy Cross 46
Collegis 44
committees 54
communicating 241
communications 136

competitive advantage 39
component-based design 107
components 103
computer-mediated conferencing 167
concerns-based adoption model 146
confidence 147
consulting 252
consumer portal 205
consumers 164
content 1
convenience 136
copyright 180
cost analysis handbook 35
costs of service delivery 30
costs of system change 30
courseware management system 84
current student inventory 34
custom channels 54
Customer Relationship Management (CRM) 202
customer satisfaction 39
customer service 193
customization 12, 80, 103
customized 69
customized view 2

D

data conferencing 167
data issues 190
database 2
Datatel 43
decision-making steps 128
defining portals 3
design consistency 20
desktop 71
destinations 2
development cycle 53
digital repositories 167
digital secretary 90, 94
Digital Teaching Assistant (TA) 94
digital tutor 94
direction and leadership 149
disintermediated 38, 104
distance learner 169
distance learning 176, 240
distributed 167
distributed learning 166

E

e-mail forwarding 195
ease of use 11, 139
eCollege CampusPortalSM 238
eCommunity 241
EDEN 108, 109
editable 247
educational debugging 33
educational purposes 29
efficiency 30
eLearning 239
employee portal 205
enterprise platform vendors 208
Enterprise Resource Planning (ERP) 63, 202
epicentric 208
evaluation 151, 163
expandability 13
expectations 144
external information 54

F

faculty inventory 35
feedback 173
feeds 2
FERPA 197
financial aid 241
first-generation portal 206
flashlight online 35
flashlight program 34
flexibility 165, 169
focus group 55, 135, 190
formative evaluation 176
Forrester Research 204
funding 188

G

Gartner 206
gateway 18, 69, 243
graphic user interface (GUI) 248

H

help desk 252
higher education 163
highlight 247
homegrown portals 42, 79

horizontal portals 42
hosted application 244
HTML 245
human-computer interfaces 163
hybrid 239

I

I-Frames 118
IBM 208
IDC 204
inactivity agent 96
inclusiveness 169
Indiana University 103
institution-wide solution 238
institutional data systems 5
institutional portal 52
instruction 29
instructional design 254
instructional technology 175
instruments 134
integration 12, 164, 169, 244
intellectual property rights 167
intelligent agent 3, 90
Intelligent portals 92
intelligent user interfaces 89
interaction 239
interface 73
international 169
Internet 73
Internet portals 1
interoperability 103
interviews 135
intra-campus bridges 1
IRISLink 187
issues 122
iterations 114

J

JA-SIG's uPortal 45
Java 217
Java 2 Enterprise Edition (J2EE) 210
Java Server Pages 209

K

Kentucky Virtual University 42
killer app 74

L

learner-centric 167
learning 167
learning communities 29
learning management 167
lessons learned 148
lifelong community 169
lifelong learning 164, 240
"living and learning" community 2

M

maintainability 11
maintenance 194
managing change 188
mediation 38
metacampus 42
Microsoft's .Net 213
middleware 109
Montana State University-Billings
    (MSU-B) 249
MSU-B Online 250
multi-channel access 217
MyLibrary 48
MyPortal 97
MyUW 44

N

"n-tier" 245
net generation 240
next generation Web portal 103
next-generation portals 90
next-generation Web sites 5
North Shore Community College 44
nuggets 245

O

oasis 217
one-stop shopping 38
one-way communication 241
OneStart 103
online study guides 243
online survey 3, 134
open architecture 115, 123
open standards 13
open-source 79

Oracle 131
Organization for the Advancement of Structured Inf 217
organizational support 151
outcomes 34

P

page design 20
partnerships 164
PeopleSoft 208
performance 13
Personal Digital Assistant (PDA) 99
personal portals 73
personalizable 2
personalization 11, 80, 103
personalized 69
physically challenged learner 169
pilot 57
planning process 128
platform compatibility 21
platform independence 13
Plumtree 208
portal 68, 241
portal platform 204
portal theory 41
portals 69
portlets 213
post-secondary 164
pre-built channels 54
Prestige 39
privacy 197
process analysis 56
productivity 5
professional accreditation 169
publishing 246
pure-play portal vendors 208

R

reach 41
recruitment 5
redesign 38
reducing costs and stresses 33
registration systems 238
reputation 39
requirements planning (MRP) 202
research 163
retention 5

richness 41
risk 196
robustness 21
role information 247
role-based authorization 20
RSS 58
Rubric for Software Selection 131

S

SAP 208
satisfaction 146
scalability 244
scenario 98
SCT 43
second-generation portal 207
security 130
selecting services 191
self-branded services 2
self-service 37
servers 196
service delivery framework 103
servlets 209
shared code approach 79
shared decision making 131
Simple Object Access Protocol (SOAP) 213
single sign-on 82, 103
single sign-on authentication 12
socialization 166
sponsors 65
staffed 188
stakeholders 137, 166
sticky 18
strategic challenges 1
strategic plan 104, 199
strategies for portal implementation 130
strategy 53
stress testing 23
student information system 78, 96, 253
students 164
study groups 243
style manager 247
surveys 191
sustainability 166
synchronous 167
system architect 16
system debugging operations 32

T

targetable 245
teaching 163
technical support 244
technology 69
technology adoption 130
Tennessee Regents Online Degree Programs 42
test environment 189
textbooks 241
the interaction 163
third-generation portal 207
timecruiser 43
tools 196
topic box 247
trainability 93
training and support 151
transformational thinking 169
transparency 164
tutoring services 243
two-way interaction 241

U

Universal Description, Discovery and Integration 217
University at Buffalo, State University of New York 44
University of Minnesota, My One Stop 44
University of Washington 44
usability 18, 115, 165
usability testing 19
usability testing methodology 22
user identity 247
user interfaces 1, 69
user-centered 242
user-centered design 105
user-defined guidelines 165
users' needs and concerns 132
user's role 3

V

vendor 73, 187
vendor solutions 130
vertical portals 42
viability 14

videoconferencing 167
virtual community 105
virtual faculty 169
Visual Editor 248
Vitria 208

W

Wake Forest University 187
Wake Information Network 187
Web index 1
Web search engine 1
Web Services 123, 216
WIN 187
wired campuses 240
wireframe 245
workflow 103
workplace learner 169

X

XML 58, 196, 247

Just Released!
The Design and Management of Effective Distance Learning Programs

Richard Discenza, University of Colorado
Caroline Howard, Emory University
Karen Schenk, K.D. Schenk and Associates Consulting

ìAnytime, anyplace, and any subjectî is an emerging theme for distance learning in higher education through out the world. Portable wireless devices and other emerging interactive media are giving traditional classroom and distance education professors a growing array of tools to provide instruction wherever it is needed or desired. Many predict that within the next year handheld devices and virtual classrooms will be ubiquitous, enabling students to log on to the Internet for assignments and to participate in chat room discussions with students across the globe. The purpose of *The Design and Management of Effective Distance Learning Programs* is to increase understanding of the major issues, challenges and solutions related to remote education. It provides the theoretical and practical knowledge of the distance education field as it currently exists in the 21st century. It addresses the technological, institutional, faculty, student and pedagogical perspectives concerning the field of distance education.

ISBN 1-930708-20-3 (h/c)ï eISBN 1-59140-001-5   US$74.95   312 pages   Copyright © 2002

> ìWith such large numbers of individuals learning at a distance from traditional educational facilities, it is critical that we understand the impacts of these arrangements, the major issues and challenges, and how to best manage distance education.î
> ñRichard Discenza, University of Colorado, USA

It's Easy to Order! Order online at www.idea-group.com or call our toll-free hotline at 1-800-345-4332!
Mon-Fri 8:30 am-5:00 pm (est) or fax 24 hours a day 717/533-8661

Idea Group Publishing
Hershey ï London □ Melbourne □ Singapore □ Beijing

An excellent addition to your library

International Journal of Distance Education Technologies (JDET)

NEW! **NEW!**

The International Source for Technological Advances in Distance Education

ISSN:	1539-3100
eISSN:	1539-3119
Subscription:	Annual fee per volume (4 issues): Individual US $85 Institutional US $185
Editors:	Shi Kuo Chang University of Pittsburgh, USA
	Timothy K. Shih Tamkang University, Taiwan

Mission

The *International Journal of Distance Education Technologies* (**JDET**) publishes original research articles of distance education four issues per year. **JDET** is a primary forum for researchers and practitioners to disseminate practical solutions to the automation of open and distance learning. The journal is targeted to academic researchers and engineers who work with distance learning programs and software systems, as well as general participants of distance education.

Coverage

Discussions of computational methods, algorithms, implemented prototype systems, and applications of open and distance learning are the focuses of this publication. Practical experiences and surveys of using distance learning systems are also welcome. Distance education technologies published in **JDET** will be divided into three categories, **Communication Technologies, Intelligent Technologies, and Educational Technologies**: new network infrastructures, real-time protocols, broadband and wireless communication tools, quality-of-services issues, multimedia streaming technology, distributed systems, mobile systems, multimedia synchronization controls, intelligent tutoring, individualized distance learning, neural network or statistical approaches to behavior analysis, automatic FAQ reply methods, copyright protection and authentification mechanisms, practical and new learning models, automatic assessment methods, effective and efficient authoring systems, and other issues of distance education.

For subscription information, contact:	For paper submission information:
Idea Group Publishing 701 E Chocolate Ave., Suite 200 Hershey PA 17033-1240, USA cust@idea-group.com URL: www.idea-group.com	Dr. Timothy Shih Tamkang University, Taiwan tshih@cs.tku.edu.tw